*To Ian Wolfgang and Megan Grace, thank you for all you have taught us and for listening when we teach you. Always be true to yourself and nice to the puppies.*

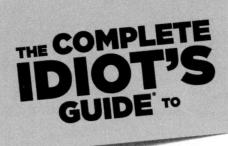

THE **COMPLETE IDIOT'S GUIDE** TO

# Electronics 101

by Sean Westcott and Jean Riescher Westcott

## ALPHA

A member of Penguin Group (USA) Inc.

## ALPHA BOOKS

Published by the Penguin Group

Penguin Group (USA) Inc., 375 Hudson Street, New York, New York 10014, USA

Penguin Group (Canada), 90 Eglinton Avenue East, Suite 700, Toronto, Ontario M4P 2Y3, Canada (a division of Pearson Penguin Canada Inc.)

Penguin Books Ltd., 80 Strand, London WC2R 0RL, England

Penguin Ireland, 25 St. Stephen's Green, Dublin 2, Ireland (a division of Penguin Books Ltd.)

Penguin Group (Australia), 250 Camberwell Road, Camberwell, Victoria 3124, Australia (a division of Pearson Australia Group Pty. Ltd.)

Penguin Books India Pvt. Ltd., 11 Community Centre, Panchsheel Park, New Delhi—110 017, India

Penguin Group (NZ), 67 Apollo Drive, Rosedale, North Shore, Auckland 1311, New Zealand (a division of Pearson New Zealand Ltd.)

Penguin Books (South Africa) (Pty.) Ltd., 24 Sturdee Avenue, Rosebank, Johannesburg 2196, South Africa

Penguin Books Ltd., Registered Offices: 80 Strand, London WC2R 0RL, England

International Standard Book Number: 978-1-61564-0-959
Library of Congress Catalog Card Number: 2010919334

13   12   11      8   7   6   5   4   3   2   1

Interpretation of the printing code: The rightmost number of the first series of numbers is the year of the book's printing; the rightmost number of the second series of numbers is the number of the book's printing. For example, a printing code of 11-1 shows that the first printing occurred in 2011.

Printed in the United States of America

**Note:** This publication contains the opinions and ideas of its authors. It is intended to provide helpful and informative material on the subject matter covered. It is sold with the understanding that the authors and publisher are not engaged in rendering professional services in the book. If the reader requires personal assistance or advice, a competent professional should be consulted.

The authors and publisher specifically disclaim any responsibility for any liability, loss, or risk, personal or otherwise, which is incurred as a consequence, directly or indirectly, of the use and application of any of the contents of this book.

Most Alpha books are available at special quantity discounts for bulk purchases for sales promotions, premiums, fund-raising, or educational use. Special books, or book excerpts, can also be created to fit specific needs.

For details, write: Special Markets, Alpha Books, 375 Hudson Street, New York, NY 10014.

**Publisher:** *Marie Butler-Knight*
**Associate Publisher/Acquiring Editor:** *Mike Sanders*
**Executive Managing Editor:** *Billy Fields*
**Development Editor:** *Jennifer Moore*
**Senior Production Editor:** *Janette Lynn*
**Copy Editor:** *Andy Saff*

**Cover Designer:** *William Thomas*
**Book Designers:** *William Thomas, Rebecca Batchelor*
**Indexer:** *Heather McNeill*
**Layout:** *Brian Massey*
**Senior Proofreader:** *Laura Caddell*

# Contents

# Introduction

The study of electronics can be a little overwhelming when you start out. But without assuming that you remember everything from your general science classes, we take you through it all step by step so that you will gain confidence in your understanding of the material. This doesn't mean that we give you a dumbed-down version of electronics, but it does mean that we cover the topics in a more digestible style. We believe that by making the effort to wrap your head around some of the harder topics, you will find it easier to progress into further study of electronic theory or hands-on experimentation.

We believe that a new revolution is under way. Electronics has always had a thriving hobbyist population, especially in the 1960s and the 1970s. There were magazines, corner electronics stores, and clubs where enthusiasts could meet and share their creations. It had its subcultures from amateur radio enthusiasts to model rocket builders. In the 1980s, this culture grew to include people building personal computers before such companies as IBM and Apple began to mass produce them.

The hobbyist field changed as electronics advanced. The increasing sophistication and miniaturization of electronic components and the products built with them made hobbyist-built electronics pale in comparison to their flashier, mass-produced competition. But those same advances are now putting the design and production back into the hobbyists' hands. Perhaps egged on by battling robots out of university engineering departments, a new generation of electronics buffs is tinkering with technology. With affordable microcontrollers and a wide range of products and information available online, the hobbyist can design and build machines that recharge the ideas of homebrew and do-it-yourself. We can all become makers.

## How This Book Is Organized

**Part 1, The Fundamentals,** covers electronics basics from the atoms up. You learn about currents, AC and DC voltage, and find out how they all work together to power our world.

**Part 2, Your Workspace and Tools,** introduces the tools of the trade, from the low-tech soldering iron that makes your connections to the high-tech digital multimeter, and offers advice for setting up a shop and working with electricity safely.

**Part 3, Electronic Components,** gives you the nitty-gritty on circuits, capacitors, diodes, transistors, and power supplies. These components are the workhorses of electronics, keeping things powered, amped up, and running smoothly.

**Part 4, Getting to Work,** keeps you busy soldering parts together and creating your own power supply. Once you have these skills under your belt, you're ready to start building—and inventing—your own electronic devices.

**Part 5, Going Digital,** teaches you to think like a computer. You learn how integrated circuits put digital signals to work and how to use memory to store the instructions that run your gadgets.

**Part 6, Constructing a Robot,** helps you use everything you learned from the previous parts to create your own robot—one that can move on its own, sense its environment, and communicate with your computer. What will you build next?

## Extras

Throughout the book, you'll find the following sidebars offering additional insights:

**DEFINITION**

Sometimes it helps to have things stated just a little more directly. In these sidebars, we save you from having to grab a dictionary.

**TITANS OF ELECTRONICS**

Not just a parade of historical figures—here we invite you to put yourself in their shoes. These sidebars offer a closer look at the folks who looked at things a little differently and changed the world with their ideas.

**HIGH VOLTAGE!**

When handled safely, electricity can be safe. But the consequences of not respecting its potential for harm are serious. The more you understand how electricity moves, the better you can prepare and work safely with it.

**WATTAGE TO THE WISE**

Here you'll find straightforward advice—sometimes practical, sometimes more philosophical.

## Acknowledgments

We would like to thank the people who helped us bring this book to publication. First, we would like to thank our agent Bob Diforio for introducing us to our fine and patient editors Mike Sanders and Jennifer Moore. Early enthusiasm and advice from André Rebelo and Extech Instruments Corporation, and the use of the Extech EX210 multimeter, were greatly appreciated. We value the advice of our technical editors, Bob Godzwon and John O'Brien. We thank the SparkFun team, including AnnDrea Boe, for helping us create supply lists and Secret Labs' Chris Walker for being excited about and supportive of our use of their microcontroller platform. Randall Monroe, of XKCD, honored us in allowing our use of his Circuit Diagram comic, we offer our thanks for his generosity. The readers and contributors to Netduino's forums provided advice in the true spirit of "free as in beer."

We would also like to thank our colleagues and especially our family for supporting us as we worked through many beautiful weekends.

## Special Thanks to the Technical Reviewer

*The Complete Idiot's Guide to Electronics 101* was reviewed by an expert who double-checked the accuracy of what you'll learn here, to help us ensure that this book gives you everything you need to know about electronics. Special thanks are extended to Bob Godzwon.

## Trademarks

All terms mentioned in this book that are known to be or are suspected of being trademarks or service marks have been appropriately capitalized. Alpha Books and Penguin Group (USA) Inc. cannot attest to the accuracy of this information. Use of a term in this book should not be regarded as affecting the validity of any trademark or service mark.

# The Fundamentals

**Part**

**1**

Electronics involves controlling the invisible. Most of the time, you see the effect of electricity but not the actual movement of electric current. This part pulls back the curtain on that hidden world to give you a peek at how electricity works at the atomic level.

It all starts with tiny, charged particles called electrons. You'll learn how and why electrons move in the natural world and how people have harnessed their power using circuits.

No overview of electronic theory would be complete without an explanation of how current (the flow of electrons, also known as electricity), voltage (the "push" that is caused by the attraction of positive to negative), and resistance (the "push back" of insulators) work. You will find out what power really means and the ways that all of these forces interact.

# The Theory Behind Electricity

## In This Chapter

- Understanding atomic structure
- Harnessing the laws of attraction and repulsion
- Controlling the flow of electrons
- Identifying an element's conductivity and resistance

Electronics is the study of devices that can control the flow of electricity. You can build devices that detect, measure, power, control, count, store, and transmit electricity—and much more. But in order to do all of these things, you first need to know what electricity is and how it flows.

To get to the essence of electricity, you must delve into some of the most basic concepts in physics: atoms and their structure.

## Atoms and Their Structure

An atom consists of a cloud of negatively charged electrons surrounding a dense nucleus that contains positively charged protons and electrically neutral neutrons. The relationship between an atom's charged particles—its protons and its electrons—is the key to electricity (much more on this in the following sections of this chapter). Atoms are basic units of matter.

Matter refers to any physical substance; in other words, matter is anything that has mass (measurable stuff) and volume (measurable occupation of space).

A chemical element is pure matter consisting of only one type of atom. Every element is composed of an atom with a particular atomic structure that defines it; for instance, the element carbon is composed exclusively of carbon atoms. Elements are ranked by their atomic number on the *periodic table of chemical elements*. The atomic number indicates the number of protons in each atom.

The standard model of an atom has an equal number of protons, neutrons, and electrons but this isn't always the case. The number of neutrons can vary, and each variation is a different isotope of that element. We call the combined number of protons and neutrons nucleons. For example, carbon-14 is an isotope of carbon. It has six protons and eight neutrons. It is still carbon, but the variation in the number of neutrons affects some of its properties.

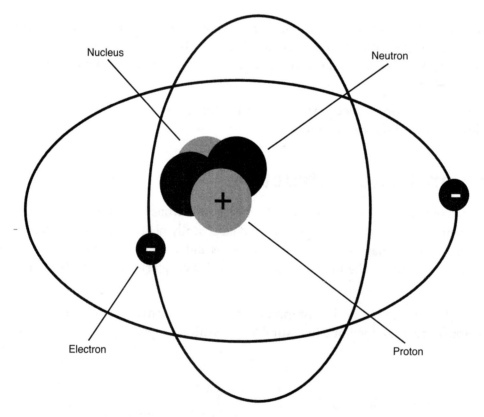

*A carbon atom.*

| 1 | 2 | 3 | 4 | 5 | 6 | 7 | 8 | 9 | 10 | 11 | 12 | 13 | 14 | 15 | 16 | 17 | 18 |
|---|---|---|---|---|---|---|---|---|---|---|---|---|---|---|---|---|---|
| hydrogen 1 **H** 1.0079 | | | | | | | | | | | | | | | | | helium 2 **He** |
| lithium 3 **Li** 6.941 | beryllium 4 **Be** 9.0122 | | | | | | | | | | | boron 5 **B** 10.811 | carbon 6 **C** 12.011 | nitrogen 7 **N** 14.007 | oxygen 8 **O** 15.999 | fluorine 9 **F** 18.899 | neon 10 **Ne** 20.180 |
| sodium 11 **Na** 22.990 | magnesium 12 **Mg** 24.305 | | | | | | | | | | | aluminium 13 **Al** 29.982 | silicon 14 **Si** 28.086 | phosphorus 15 **P** 30.974 | sulfer 16 **S** 32.065 | chlorine 17 **Cl** 35.453 | argon 18 **Ar** 39.948 |
| potassium 19 **K** 39.098 | calcium 20 **Ca** 40.078 | scandium 21 **Sc** 44.956 | titanium 22 **Ti** 47.867 | vanadium 23 **V** 50.942 | chromium 24 **Cr** 51.996 | manganese 25 **Mn** 54.938 | iron 26 **Fe** 55.845 | cobalt 27 **Co** 58.933 | nickel 28 **Ni** 58.693 | copper 29 **Cu** 63.546 | zinc 30 **Zn** 65.39 | gallium 31 **Ga** 69.723 | germanium 32 **Ge** 72.61 | arsenic 33 **As** 74.922 | selenium 34 **Se** 78.96 | bromine 35 **Br** 79.904 | krypton 36 **Kr** 83.80 |
| rubidum 37 **Rb** 85.486 | strontium 38 **Sr** 87.62 | yttrium 39 **Y** 88.906 | zirconium 40 **Zr** 91.224 | niobium 41 **Nb** 92.906 | molybdenum 42 **Mo** 95.94 | technetium 43 **Tc** 98 | ruthenium 44 **Ru** 101.07 | rhodium 45 **Rh** 102.91 | palladium 46 **Pd** 106.42 | silver 47 **Ag** 107.87 | cadmium 48 **Cd** 112.41 | indium 49 **In** 114.82 | tin 50 **Sn** 118.71 | antimony 51 **Sb** 121.76 | tellurium 52 **Te** 127.60 | iodine 53 **I** 126.90 | xenon 54 **Xe** 131.29 |
| caesium 55 **Cs** 132.91 | barium 56 **Ba** 137.33 | lutetium 71 **Lu** 174.97 | hafnium 72 **Hf** 178.49 | tantalum 73 **Ta** 180.95 | tungsten 74 **W** 183.94 | rhenium 75 **Re** 186.21 | osmium 76 **Os** 190.23 | iridium 77 **Ir** 192.22 | platinum 78 **Pt** 195.08 | gold 79 **Au** 196.97 | mercury 80 **Hg** 200.59 | thallium 81 **Tl** 204.38 | lead 82 **Pb** 207.2 | bismuth 83 **Bi** 208.98 | polonium 84 **Po** 209 | astatine 85 **At** 210 | radon 86 **Rn** 222 |
| francium 87 **Fr** 223 | radium 88 **Ra** 226 | lawrencium 103 **Lr** 262 | rutherfordium 104 **Rf** 261 | dubnium 105 **Db** 262 | seaborgium 106 **Sg** 266 | bohrium 107 **Bh** 264 | hassium 108 **Hs** 269 | meitnerium 109 **Mt** 298 | **?** | **?** | **?** | | | | | | |

| lanthanum 57 **La** 138.91 | cerium 58 **Ce** 140.12 | praseodymium 59 **Pr** 140.91 | neodymium 60 **Nd** 144.22 | promethium 61 **Pm** 145 | samarium 62 **Sm** 150.36 | europium 63 **Eu** 151.96 | gadolinium 64 **Gd** 157.25 | terbium 65 **Tb** 158.93 | dysprosium 66 **Dy** 162.50 | holmium 67 **Ho** 164.93 | erbium 68 **Er** 167.26 | thulium 69 **Tm** 168.93 | ytterbium 70 **Yb** 173.04 |
|---|---|---|---|---|---|---|---|---|---|---|---|---|---|
| actinium 89 **Ac** 227 | thorium 90 **Th** 232.04 | protactinium 91 **Pa** 231.04 | uranium 92 **U** 238.03 | neptunium 93 **Np** 237 | plutonium 94 **Pu** 244 | amercium 95 **Am** 243 | curium 96 **Cm** 247 | berkelium 97 **Bk** 247 | californium 98 **Cf** 251 | einsteinium 99 **Es** 252 | fermium 100 **Fm** 257 | mendelevium 101 **Md** 258 | nobelium 102 **No** 259 |

*The periodic table of elements.*

**DEFINITION**

The **periodic table of chemical elements,** often simply called the *periodic table,* lists the 118 known elements and basic information—atomic number, relative atomic mass (also known as atomic weight), symbol, and other information, depending on the table—about each element.

# Electrons

The atomic number of an element indicates the number of protons. For an electrically neutral or stable atom the number of protons and electrons are equal, which means that once you know the atomic number of an element you know the number of electrons it has. Electrons travel around the nucleus of the atom in an area known as a *shell*. Shells are layered outward from the nucleus. Each shell can hold up to a maximum number of electrons.

The innermost shell can hold 2 electrons, the second shell can hold 8, the third shell can hold 18, and the fourth can hold 32.

The following table shows the electron arrangements for some common elements:

## Number of Electrons in Various Elements

| Atomic Number | Element Name | Electrons per Shell |
|---|---|---|
| 1 | Hydrogen | 1 |
| 2 | Helium | 2 |
| 6 | Carbon | 2,4 |
| 7 | Nitrogen | 2,5 |
| 8 | Oxygen | 2,6 |
| 9 | Fluorine | 2,7 |
| 10 | Neon | 2,8 |
| 11 | Sodium | 2,8,1 |
| 12 | Magnesium | 2,8,2 |
| 13 | Aluminum | 2,8,3 |
| 14 | Silicon | 2,8,4 |
| 15 | Phosphorus | 2,8,5 |
| 16 | Sulfur | 2,8,6 |
| 17 | Chlorine | 2,8,7 |
| 19 | Potassium | 2,8,8,1 |
| 20 | Calcium | 2,8,8,2 |
| 24 | Chromium | 2,8,13,1 |
| 26 | Iron | 2,8,14,2 |
| 28 | Nickel | 2,8,16,2 |
| 29 | Copper | 2,8,18,1 |
| 30 | Zinc | 2,8,18,2 |
| 33 | Arsenic | 2,8,18,5 |
| 36 | Krypton | 2,8,18,8 |
| 47 | Silver | 2,8,18,18,1 |
| 50 | Tin | 2,8,18,18,4 |
| 53 | Iodine | 2,8,18,18,7 |
| 79 | Gold | 2,8,18,32,18,1 |
| 80 | Mercury | 2,8,18,32,18,2 |
| 82 | Lead | 2,8,18,32,18,4 |
| 92 | Uranium | 2,8,18,32,18,8,2 |

# Valence Shell

The outermost shell of an atom is known as the *valence shell* (or *valence band*), and the electrons that inhabit that outer shell are called *valence electrons.* The more full the valence shell, the less likely it is that an atom will lose electrons when a force is applied. The less full the valence shell, the more likely it is to lose electrons when a force is applied.

Let's compare two elements. As you can see from the preceding table, neon has a full valence shell, meaning that it is unlikely to gain or lose electrons. Copper, on the other hand, has just 1 electron in its valence band, which can hold 32 electrons. This lone electron filling the valence shell is easily attracted away to a nearby atom that has room on its valence shell.

If a valence shell loses or gains an electron, the atom becomes an *ion.* An ion is an atom with a charge. An atom that has more protons than electrons has a positive charge. An atom with more electrons than protons has a negative charge. Because of *electromagnetic force*, negatively charged electrons will leave their own valence shell to travel to another atom that has a positive charge.

**DEFINITION**

**Electromagnetic force** is the attraction between positive and negative charges and the repulsion of like charges. It is the basis of interaction between the protons and electrons within atoms holding them together, and the attraction between atoms that have negative and positive charges.

Here's where electricity enters the picture: The movement of electrons on the valence shell when leaving or joining another atom creates electrical current, or electricity. The movement of electrons (and therefore electricity) relies on the two basic concepts that result from electromagnetic force: 1) Opposite charges are attracted to each other; and 2) like charges repel each other.

**WATTAGE TO THE WISE**

In his pioneering work on electricity, Benjamin Franklin described something that produced electricity as *positive* (positive because it gave current) and the recipient material of that current as *negative* (because it was receiving the electrical charge). This is called *conventional theory*—the early belief that current traveled from positive to negative.

Today we now know that the opposite is true: Current travels from negative to positive. This is called *electron theory.*

Confusingly enough, many diagrams that are used to describe circuits show the flow of current in *conventional notation*, with current flowing from the positive terminal of a battery to the negative terminal. Others use the more accurate *electron notation*. History creates traditions; we sometimes have to learn to go with the flow!

# Conductors, Insulators, and Semiconductors

Some atoms are more stable (or neutral) than others. Stable atoms have an equal number of positively charged protons and negatively charged electrons. The attraction between protons with positive charges and electrons with negative charges holds the atom together unless a force is introduced to separate them.

*Conductivity* is the tendency of a material to allow the free flow of electrons. *Resistance* is the opposite; it is the tendency of a material to resist the flow of electrons. When we measure conductivity, we refer to it as resistance. A good conductive material is simply said to have very low resistance. The conductivity of a material is determined by how full or empty the valence shell of its atoms is.

An atom with a full valence shell is not going to accept extra electrons, while an atom with a nearly empty valence shell will be able to shed and receive electrons. This flow of electrons among atoms is electricity. As we mentioned previously, copper's nearly empty valence shell allows it to shed and accept electrons, so it is a good conductor of electricity. Neon, with its full valence band, is very nonreactive, so it is resistant to the flow of electricity; in other words, it is an insulator.

Knowing both of these qualities is important to understanding electronics. Electronics relies on our having the ability to control the flow of electricity. We need to be able to slow it, block it, and even modulate it. (More on that later!) This requires that we understand which materials are conductors (highly conductive or low resistance), which are insulators (poor conductors or strong resistance), and which are semiconductors (in between low and strong resistance).

## Conductors

Elements that are grouped on the left side of the periodic table have fewer electrons in their valence shell and can serve as good conductors. That's because these electrons are loosely bound to their nuclei (the plural of *nucleus*) and can easily be separated from their atom and travel to a positively charged ion. In other words, these elements allow for electricity—which is simply the flow of electrons—to flow easily. Examples of common metals that are relatively good conductive materials are silver (Ag), gold (Au), and copper (Cu), all of which contain just one electron in their valence shell; that lone electron is easily removed when electricity is flowing. Moving to the right from these metals to the far right of the periodic table you encounter more stable elements that are less conductive.

## Insulators

Elements that have full or nearly full valence shells either hold on to their existing electrons or attract electrons so that their valence shell becomes full. These elements are insulators that have great resistance and can slow or block the flow of electricity. They don't have room on their valence shell to accept electrons, and their nearly full outer shell holds tightly to the electrons it already has.

The elements at the far right of the periodic table are called the noble gases. These are extremely good insulators as they are very nonreactive. The naturally occurring noble gases are helium (He), neon (Ne), argon (Ar), krypton (Kr), xenon (Xe), and radon (Rn).

## Semiconductors

The elements in between the metals and the noble gases on the periodic table are generally semiconductors. Some elements commonly used as semiconductors are silicon (Si) and germanium (Ge). These elements can be combined with others to introduce impurities that can conduct electricity. This process is called *doping*, and when an element is used in this capacity, it is referred to as a *dopant*.

Consider an atom of silicon, which has four electrons in its valence shell. When you look at multiple atoms of silicon, as shown in the following figure, you can see that they arrange themselves quite neatly into what is called a *crystalline structure*, meaning the atoms form a repeating pattern in each direction, with each of its electrons in the valence shell perfectly paired with its neighboring atom.

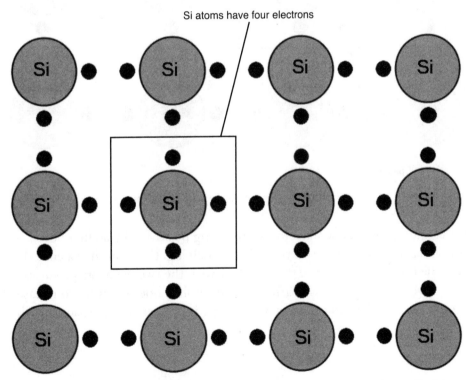

*Atoms of silicon in a crystalline structure. Note that the image doesn't depict all of the atoms on the outside rows; silicon has four electrons in its valence shell.*

Now let's look at what happens when two very different dopants are added. Phosphorus (P) has 15 electrons—2 on its inner shell, 8 on its second shell, and 5 on its valence shell. When it bonds with silicon, the combination yields a loosely attached electron. Because that electron can be easily released, a negative charge can easily flow through the doped semiconductor. Phosphorus acts as a donor impurity, because when it is added to silicon it releases or donates electrons. This yields what is called an *n-type* semiconductor, where *n* means negative.

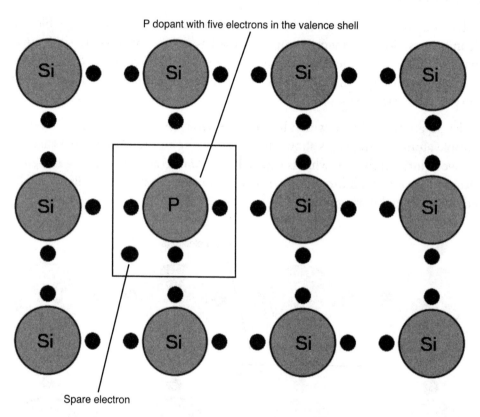

*An n-type semiconductor formed by doping with phosphorus.*

You can create a *p-type* semiconductor—*p* meaning positive—by adding boron (B) to silicon. Boron has five electrons, two on its inner shell and three on the valence shell. When you combine these two elements, the bond between the two elements produces a valence shell with seven electrons. This nearly full valence shell does not want to release electrons. However, it does have room to accept an electron in the remaining space, which is referred to as a *hole*.

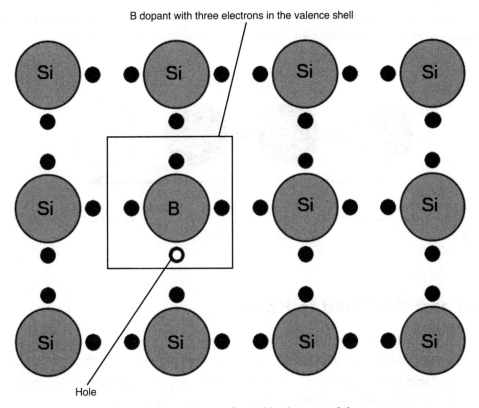

*A p-type semiconductor formed by doping with boron.*

**WATTAGE TO THE WISE**

Although the atomic qualities of a material are the most important in determining its conductivity, other factors need to be considered when determining an element's conductivity:

- **The physical characteristics of the material:** A thick strip of aluminum will conduct more electricity than a thin one. A short wire shortens the distance needed for the current to travel compared to a longer wire.

- **Temperature:** Different materials change in their conductivity depending on the temperature. Metals tend to become less conductive when heated and some become superconductive at extremely low temperatures.

## Electron Flow Versus Hole Flow

When an electron leaves an atom, it creates a gap for the next electron to jump into. The electrons move in one direction, so the gaps always open up in the reverse direction.

The movement of the electrons is called *electron flow*. The opening up of the gaps is called *hole flow*.

The flow of electrons is like a flow of marbles through a straw. One electron moves into the space created by the movement of the previous electron down the line.

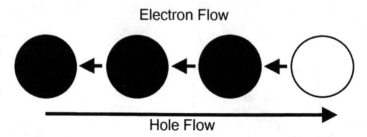

*When electrons move to the empty holes during electron flow, the "movement" of the empty holes is called hole flow.*

## The Least You Need to Know

- An atom has a nucleus containing positively charged protons and neutrally charged neutrons; the nucleus is surrounded by a cloud of negatively charged electrons.
- A stable atom has equal numbers of protons and electrons. When an atom gains or loses an electron, it becomes an ion—a charged atom.
- Electrons travel in shells or bands around the nucleus. The outer shell is called the valence shell. When electrons move from one atom to another, they create an electric current.
- Materials are classified as conductors, insulators, or semiconductors based on their resistance to conductivity. Semiconductors can be doped to create n-type or p-type semiconductors.

# How Electricity Works

## In This Chapter

- Creating paths for electricity to follow
- Giving electricity a push
- Measuring voltage, current, and resistance
- Calculating power using Ohm's Law and Joule's Law

Now that you know what electricity is at the most fundamental level, it's time to find out more about how it flows and how you can take charge of that flow. Electricity needs a path and a push. Once you understand how to manipulate the path and the push, you can control the devices you connect to the path.

## Circuits

The path on which electricity flows is called a *circuit*. Once flow has been established, electric current can travel endlessly through a conductive material if the circuit remains as a loop. Chapter 1 compared the flow of electric current to marbles moving through a looped straw. In this comparison, the circuit is the straw, and it can't carry electricity if there is a break anywhere along it.

More practically, a circuit is any arrangement that allows for electrical current to flow. An example of a very basic circuit might be a battery connected to a lamp. A computer's motherboard contains several much more complicated circuits. Electronics is all about analyzing, building, and creating circuits that use electrical current.

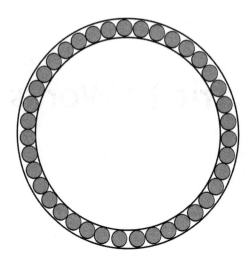

*A circuit is like marbles moving through a looped straw.*

The circuit's current provides power for a device or devices. The device that is powered by a circuit is called the *load*. Wire connects the battery and the load. In the basic circuit of a lamp and a battery, the lamp is the load. This basic circuit consists of a power supply (the battery), a load (the lamp), and the wire.

# Electromotive Force or Voltage

How does the flow get started? *Electromotive force* (you may see it abbreviated as emf in discussions about electricity and represented by the symbol V in equations) is the push that gets the electrons jumping from one atom to another, sending a current of electrical flow along the way. Electromotive force can originate from many sources, including the following:

- Chemical reactions, as in a battery

- Electromagnetic generators

- Photovoltaic cells (solar)

- Generators that convert mechanical energy to electrical energy

- Friction

- Thermoelectrical sources, which use differences in temperature to create electricity

Voltage (V), named after Italian scientist Allessandro Volta, is the measurement of emf. It is the measure of the force required to move electricity between two points on a circuit, known as the potential difference (p.d.) between those two points. You cannot measure the voltage at a single point; it is always a measurement across two points. Like speed or length, to measure voltage you need to have two points to show a relationship.

It may seem that we are using a lot of words to represent the same concept and, frankly, we are. Hopefully, using them all in the same sentence will help clarify relationships: Electromotive force (emf), also known as voltage (V), is the potential difference between two points in a circuit; it is symbolized by the letter E and is measured in volts.

**TITANS OF ELECTRONICS**

The first practical electrical generator was designed by Michael Faraday in 1831. He discovered that if you rotate a conductive metal wire in a magnetic field, a process called induction (see Chapter 11), you can generate a current. A generator uses mechanical energy to turn the wire, converting mechanical energy into electric energy that can cause current to flow through a circuit.

Some generators are hydroelectric, meaning that they use the flow of water to turn a turbine. Oil and coal can be burned to cause steam, which also turns a turbine to generate electricity. Atomic energy uses the heat released by nuclear fission to create steam to turn a turbine. Even green energy such as wind power relies on a turning turbine to create electricity through induction.

# Current

Because current (I) is all electrons moving through a circuit, we measure it as it moves through a single point. To account for the fact that electrons are incredibly small, a large unit was created to represent a set number of electrons. A *coulomb* (pronounced *KOO-lum*) is equal to approximately $6.25 \times 10^{18}$ electrons. An *ampere* (pronounced *AM-peer* and abbreviated as amp or simply A) is defined as a coulomb of current that moves through a point in one second.

**TITANS OF ELECTRONICS**

French physicist Charles Augustin de Coulomb is the namesake of the coulomb.

The ampere is named after French scientist André-Marie Ampère.

Again, let's put all these terms together in a single sentence: Current is the number of electrons that move in a circuit, it is symbolized as I, and is measured in a unit called an ampere (A).

# Resistance

Resistance is the oppositional force to emf. It might help to think of resistance as the equivalent of friction slowing down a moving object. As current is pushed through a circuit by voltage, it encounters resistance, which reduces the voltage. This is why we measure voltage across different points along the circuit. The resistance of the material that makes up the circuit determines how much the voltage is reduced.

We measure resistance in relationship to voltage and current. An ohm is the level of resistance that allows the one volt of emf to move one ampere across two points on a circuit. The symbol for an ohm is the Greek letter omega, $\Omega$.

See the following table for a review of basic electronic measurements.

**Electronic Measurements**

| Quantity | Symbol | Unit of Measurement | Unit Abbreviation |
| --- | --- | --- | --- |
| Current | I | Ampere (amp) | A |
| Voltage | E or V | Volt | V |
| Resistance | R | Ohm | $\Omega$ |

# Ohm's Law

You have now learned the three measurements that are a part of the most basic formula in the field of electronics: Ohm's Law. Ohm's Law states that the current (I) between two points is directly proportional to the voltage (V) and inversely proportional to the resistance (R). As an equation, it is written I = V/R. If you have any two of the variables, you can solve for the other. For instance, if you have R and I, you can solve for V using this equation: V = R × I. Similarly, if you know the values of V and I, you can solve for R with this equation: R = V/I.

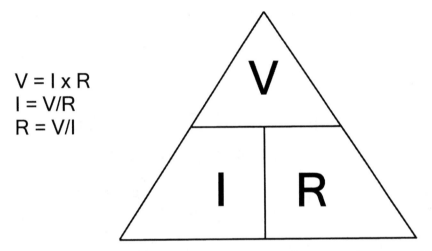

$$V = I \times R$$
$$I = V/R$$
$$R = V/I$$

*Ohm's Law pyramid shows the relationship between voltage, current, and resistance. Note that if you look at any of the segments of the pyramid, the other two values are shown in their mathematical relationship.*

# Power

In a direct current circuit (see Chapter 2 for more on direct current), *power* is voltage multiplied by current. The unit of measurement for power is the watt (W), named after the Scottish scientist James Watt. One volt pushing one amp of current equals one watt.

You may be more familiar with the term kilowatt (kW) as a unit of power. A kilowatt is 1,000 watts. Your electric bill lists the number of kilowatt hours (kW-h)—the amount of power in total when a kilowatt of power is delivered constantly over an hour—you use each month. The average American home uses about 700 kW-h a month. To put this in some context, think of a 50 watt bulb. If you use that bulb for one hour, you have used 50 Wh, (watt hours) and if you use it for 20 hours, you will have used 1 kW-h because 50 Wh × 20 hours equals 1,000 Wh or 1 kW-h.

# Joule's Law

Ohm's Law shows the relationship between current, voltage, and resistance. If you want to determine power you need to know another foundational law of electricity: Joule's Law. You can use Joule's Law to calculate the amount of power provided by a circuit. Joule's First Law gives us the following equation:

power = voltage × current

**TITONS OF ELECTRONICS**

Joule's Law is named after James Prescott Joule, a British physicist and brewer.

You can combine Joule's Law and Ohm's Law to solve for voltage, current, resistance, and power.

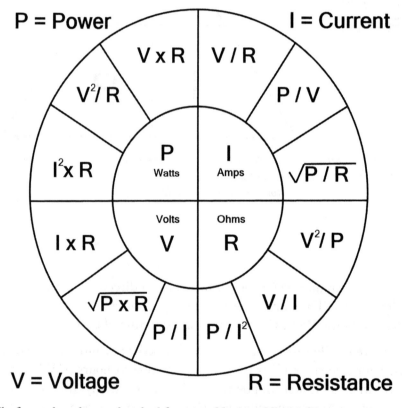

*The four values that can be solved for using Ohm's and Joule's Law: power, current, voltage, and resistance. In each quarter of the circle you can see the variables needed and the relevant equation to solve for each value.*

# Putting It All Together

To help understand the flow of electricity, imagine a football player running down a field facing a team of defenders trying to prevent him from going forward. The football player is a direct current, and the field is a circuit. The speed the player runs at is the voltage. The defenders on the field represent the resistance of the circuit.

As the player encounters the resistance of the defenders, his speed (voltage) decreases. If the resistance is small, the player (the direct current) can move through easily, but if the resistance is large, his voltage will decrease more quickly. To determine the player's voltage, you can take measurements across the yard line markers to show the effect of the resistance on him.

The size of the player—his current multiplied by his voltage—determines the power he delivers. His power will change across the circuit because the resistance will reduce his voltage. A large current with little resistance will have a lot of power; a small current with greater resistance will have significantly less power.

The player's size remains constant, so the current he represents can be measured anywhere in the circuit. Resistance doesn't affect the amount of current. The defenders' size or resistance can also be measured at any point because they are stationary.

## The Least You Need to Know

- Voltage equals electromotive force, which equals potential difference (p.d.). It is represented by the letter V and is measured in volts (V).
- Current is the measure of the flow of electrons. It is represented by the letter I and is measured in amps (A).
- Resistance is the oppositional force to flow in a circuit. It is represented by the letter R and is measured in ohms ($\Omega$).
- Power is the combination of voltage and current. It is measured in watts (W).
- Ohm's Law says that $V = I \times R$.
- Joule's First Law says that $P = V \times I$.

## Lab 2.1: Constructing a Simple Circuit

In constructing this simple circuit you can see the components of a circuit in action.

**Materials:**

> Flashlight bulb (lamp)
>
> 9 V battery
>
> Jumper wire
>
> Masking tape

**Instructions:**

Connect the battery with the jumper wires using masking tape as shown in the diagram. Be sure to connect the wire from the positive terminal to the bottom of the lamp bulb base. Connect the wire from the negative terminal to the side of the bulb base using tape.

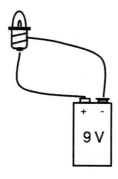

# Currents and Circuits

## In This Chapter

- Introducing direct current (DC) and alternating current (AC)
- Analyzing currents by the waveforms they produce
- Measuring voltage in AC
- Identifying different types of circuits

Electrical current can travel as a direct current (DC) and an alternating current (AC). Both types of current are the movement of electrons, and both can produce power, but they are not interchangeable. Most basic electronics encountered by hobbyists or students are powered by direct current (DC). When working with electronics we are primarily concerned with direct current, but it is important to be familiar with the properties of each.

## Direct Current

Direct current (DC) is current that travels in one direction only. Batteries of all sorts provide DC power. In a DC circuit, the current flows from the negative terminal of the battery through the circuit to the positive terminal of the battery.

DC is also said to have constant *polarity*, meaning that it doesn't change from positive to negative or vice versa; it remains as either one or the other. This will become important to know as you learn more about electronics.

# Alternating Current

Alternating current (AC) has electron flow in both directions, alternating between forward and backward. It also has changes in polarity.

Most households receive AC current from their power company. AC is easier to transmit at large voltages across greater distances to neighborhood substations. Transformers are used at these substations to lower the voltages for levels appropriate for household use. You'll learn more about working with AC as a power source later in this book, but knowing the basics will equip you with all you need to get started working safely.

# The War of Currents

The decision to use AC power to bring electricity to American homes was not made easily. Inventor Thomas Edison believed strongly in DC being the safest technology. Of course, he was significantly invested in DC infrastructure: if his DC generation and delivery systems had been widely adopted, he would have become a very rich man.

**TITANS OF ELECTRONICS**

Thomas Alva Edison (1847–1931) is one of America's greatest inventors and scientists. His inventions changed the way we live our lives, from electricity in the home to mass communication and entertainment. His early successes with inventions like the quadroplex telegraph (which allowed for more than four signals to travel on the same line) allowed him to finance his great industrial laboratory at Menlo Park, New Jersey. His legacy includes inventions or improvements in the following technologies: incandescent lightbulb, the phonograph, kinetoscope, stock ticker, film projectors, and the delivery of electricity to homes and businesses.

While Edison was refining DC generation, George Westinghouse and Nikola Tesla were working together on AC generation and transmission. They believed that AC's ability to be transmitted over much farther distances made it the clear choice, as fewer generation stations would be needed.

Edison went on a publicity campaign trying to prove that AC was dangerous because it used much higher voltages. He arranged to electrocute stray animals to death and built New York's electric chair to show the public the dangers of high voltage. He used the term "Westinghoused" instead of electrocution—which obviously didn't catch on.

Some communities did use DC as their current of choice. Edison's electric company (later folded into the New York utility Consolidated Edison, or Con Edison) provided power to areas in New York City and Westchester County, and other DC systems powered parts of Boston. DC was phased out slowly starting in the 1960s, and the last of Con Edison's DC-powered homes were converted to AC in 2007.

DC still does provide electricity in some situations. Many homes that go "off the grid" and are self-sufficient store their locally generated power in DC batteries. Some developing countries use high-voltage DC transmission systems, and telecommunications are often powered by DC.

# Waveforms

Steady DC produces a constant voltage between two points without any additional resistance. AC voltage constantly varies among peak positively charged voltage, no voltage, and peak negatively charged voltage. It does this in a regular pattern, and the varying voltage can be described by classifying its waveform.

When you look at a waveform, you can identify a few of the parts:

- **Amplitude** is the measurement of the distance of any point of the wave that is above or below the center or mean line.

- The **peak amplitude** is the point of the wave that is farthest from the mean line; it can be positive (above the mean line) or negative (below the mean line).

- The **cycle** of a wave is one complete evolution of its shape.

- The **period** is the amount of time it takes to complete one cycle. The symbol for the period is T and it is measured in seconds(s) or milliseconds (ms).

- The **frequency** is the number of complete cycles in a given amount of time. The symbol for frequency is f. Frequency is measured in Hertz (Hz); it is named after the German physicist Heinrich Hertz. One Hz is equal to one cycle per second. Most U.S. AC power is delivered at 60 Hz, meaning 60 cycles between negative and positive and back to the center line again in one second. Frequency and period are reciprocal to each other. Their relationship can be expressed as $f = 1/T$ or $T = 1/f$.

**WATTAGE TO THE WISE**

Two examples of specialized oscilloscopes are the electrocardiograph (EKG) machines, which allow medical personnel to analyze the waveform of your heartbeat, and fetal monitors, which track uterine contractions during labor. Both devices depict the waveforms so they can be monitored for abnormalities.

The best way to "see" AC waveforms is to use an oscilloscope, a professional tool that depicts the waveforms produced by electrical current. It allows you to quickly see the voltage and wave characteristics of the current.

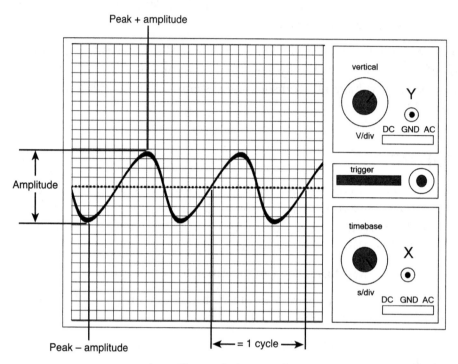

*An oscilloscope depicts waveforms.*

## Sine Wave

To study waveforms further, let's look at an AC generator, called an alternator. On this type of electrical generator, as the rotating magnetic core (the rotor) rotates within a stationary wire (the stator) the polarity of the charge and the direction of the current change.

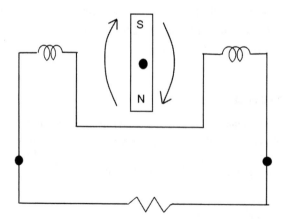

*An AC generator, called an alternator. At this point there is no current flow.*

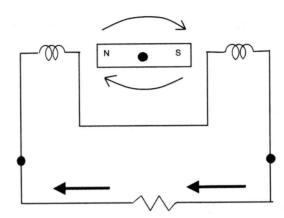

*As the rotor starts its rotation, a current is generated and the current flows in a negative direction.*

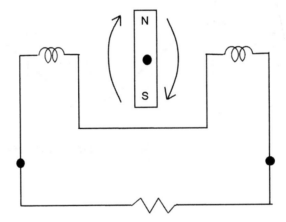

*As the rotor returns to a vertical position there is no current flow.*

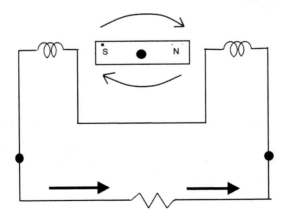

*As the rotor turns again, current flows in a positive direction.*

In this constant, smooth rotation, the current flows in a continuous *sine wave* shape as the voltage fluctuates between a positive and negative charge.

**DEFINITION**

A **sine wave** (also called a *sinusoidal wave*) is the shape that results from plotting the mathematical equation $y = \sin(x)$. The sine wave is the shape that occurs most often in ocean waves, sound waves, and light waves.

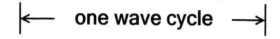

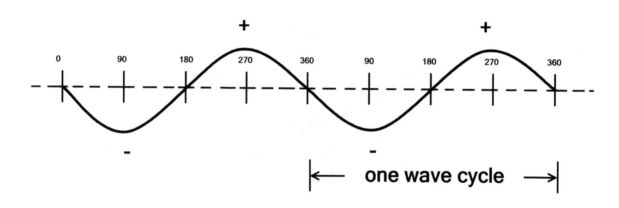

*A sine wave shape depicting current produced by a rotating alternator shaft.*

## Other Waveforms

Sine waves are not the only waveforms that occur in electronics. Square, sawtooth, and triangle waves are some of the most common regular nonsinusoidal waveforms. There are also irregular forms of both the sine waveforms and these other shapes, but for our purposes, we need only identify the general shape of the waveform.

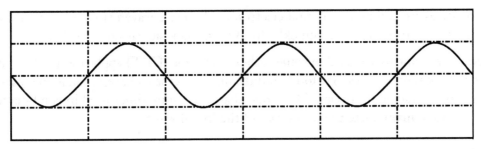

*Sine waveforms.*

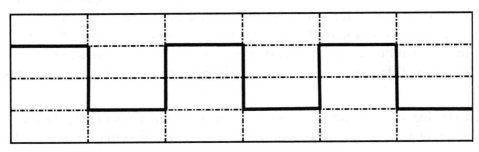

*Square waveforms.*

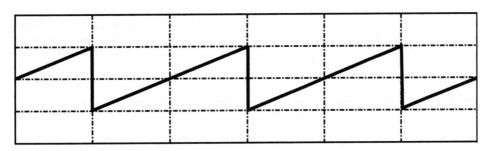

*Sawtooth waveforms.*

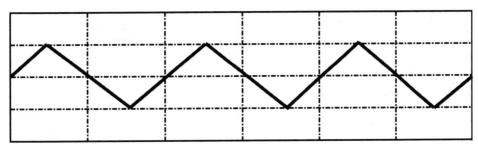

*Triangle waveforms.*

Square waves are common in digital electronics. A square waveform depicts a current that sustains a certain voltage, then quickly drops to an opposite sustained voltage.

Sawtooth waveforms are used in some cathode ray tube (CRT) applications and in manipulations of vocal harmonics. Sawtooth and triangle waveforms have gradual increases in voltage, then gradual decreases. These odd shapes are sometimes used in combination with other waves to manipulate the properties of the initial wave.

## Phase

In electronics, *phase* is the relationship between two waves. Two overlapping waveforms are said to be "in phase." If two waveforms are of the same frequency and voltage but don't overlap, they have a *phase shift*. If you think of two runners on a race track starting at the same time from different points, or at staggered start times, they can be said to be in phase shift. The time difference or space difference between the runners is the amount of phase shift.

When working with AC, you would apply trigonometric analysis to describe the wave and the amount of phase shift. The sine wave is representative of a circle's rotation. We can look at the points along the timeline of a waveform as differences in degrees of rotation. Phase shift can thereby be described as being *x* degrees out of phase.

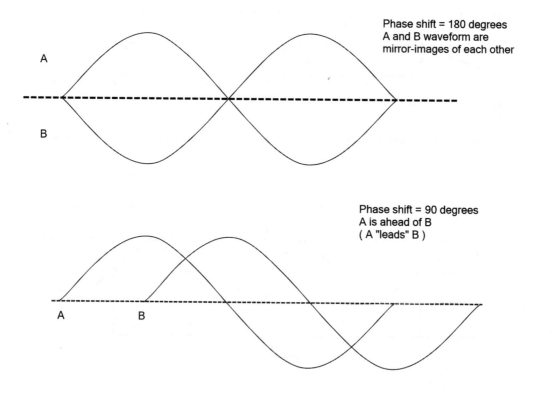

Phase shift = 180 degrees
A and B waveform are
mirror-images of each other

Phase shift = 90 degrees
A is ahead of B
( A "leads" B )

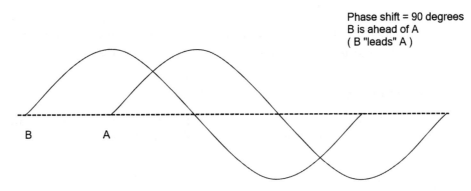

Phase shift = 90 degrees
B is ahead of A
( B "leads" A )

*Out-of-phase sine waveforms.*

When two waveforms are out of phase, at any single point you can measure the phase difference between the two. The first wave is called the leading wave and the next wave is the lagging wave. The amount of phase shift can be measured across the time axis.

If two waveforms of the same frequency and voltage have a phase shift of 180 degrees, they cancel each other out because they have opposite voltages at every point in their waveform except for when they meet at the center or mean line.

## Using Waves to Measure AC Voltage

AC is constantly fluctuating, so how do we determine the voltage? One measure is *peak positive voltage* when the positive charged voltage is at its highest point above the mean line. Another measure is *peak negative voltage*, when the voltage is at the lowest point below the mean line. This is an important measure because you will want to ensure that any circuits you design can tolerate the peak voltages. Of course, the peak voltages are produced in AC only at the top and bottom of the waves, so a circuit that requires a steady flow of those voltages would require a different measurement. Peak voltage is indicated by a subscript P or PK, as in 200 $V_p$ or 200 $V_{pk}$.

Another useful measure is *peak-to-peak voltage*. It is the combined amplitudes from peak positive to peak negative. It is abbreviated P-P. P-P voltage values also use a subscript, this time PP, so 200 $V_{pp}$.

How do you go about finding the average voltage? If you average the values of all the points along the wave, the result is always zero, as there are always just as many points above the mean line than below it. Obviously, there is more voltage in AC flow than zero. However, it is sometimes useful to know the average of just one half of the cycle: the positive or negative values averaged. We could use calculus to determine the *average value* of just each half of the cycle, but some shapes have a known value. For a sine wave, you can multiply the peak voltage times .636 to get the average value. This is expressed as $V_p(.636)$.

To give a more workable voltage of AC flow, use the concept of *root-mean-square* (rms) voltage or effective voltage. This is often given as DC equivalent. For a sine waveform, the rms can be calculated by multiplying the peak voltage times .707 or $V_p(.707)$.

**WATTAGE TO THE WISE**

Oscilloscopes, voltmeters, or multimeters (covered in Chapter 4) are the tools that most practitioners use to get accurate measurements of the key values of AC or DC.

## Direct Current Waveforms

The electronics projects covered in this book use DC. The power source could be a battery or AC power supply converted to DC through a device called a *rectifier*. (An inverter converts DC to AC.)

Remember that DC is a unidirectional flow of electrons as opposed to the alternating directions of flow in AC. In a circuit with a constant voltage, the waveform for the DC is a straight line horizontally over time. Decreasing voltage due to resistance or components that increase or decrease the voltage will make the line vary in amplitude, but it will never change direction.

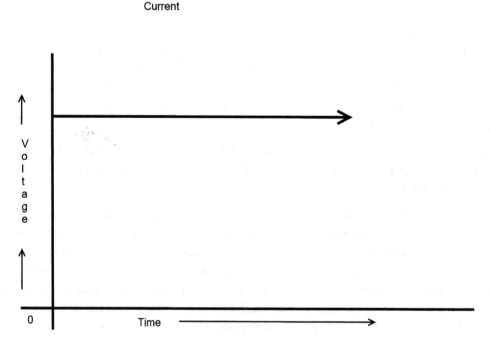

*Waveform of DC with constant voltage.*

# More About Circuits

Now that you are a little more versed in the differences between DC and AC and the waveforms that they create, let's take a closer look at circuits, or the paths through which current travels.

You already know that a circuit in electronics is a continuous loop of conductive material that allows for the flow of electrons. One of the most basic circuits is a battery connected to a lamp by a copper wire. The flow across the circuit is from the negative terminal of the battery through the lamp (providing the power for the lamp to light) and then to the positive terminal of the battery.

## Circuit Diagrams

In electronics, diagrams are used to describe circuits. These circuit diagrams depict the components added along the path of the electron flow.

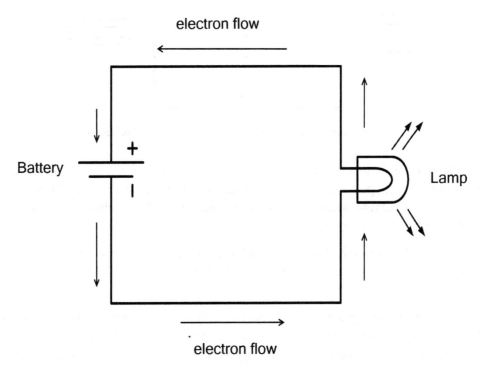

*Simple circuit diagram showing a battery and lamp.*

Most circuits contain a voltage source (a battery in our example), a path (conductive wire), and a load (the component that does the work of the circuit; in this circuit, the lamp is the load). The circuit diagram depicts these using simple symbols for each part of the circuit.

Closed

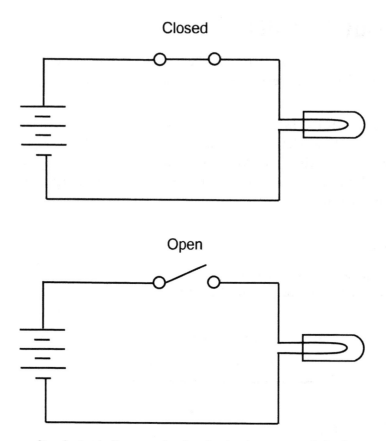

Open

*Simple circuit diagrams showing the circuit as open and closed.*

A closed circuit provides constant flow of current. If there are any breaks in the circuit—anything that interrupts current flow—it is an open circuit, also called a broken circuit. In open circuits, current doesn't flow at all, because a break anywhere along the path stops current flow from any part of the circuit.

Most circuits have a switch. A simple switch causes a break in the circuit to stop the flow of current. In our example, we could add a switch that would serve as an on/off switch for the lamp.

When making or reading a circuit diagram, it is important to pay attention to the polarity of the voltage source's terminals. Remember that the polarity reflects the charge of the terminal, positive (+) or negative (–). On a typical battery, the terminals are labeled clearly. Your circuit diagram should be labeled as well.

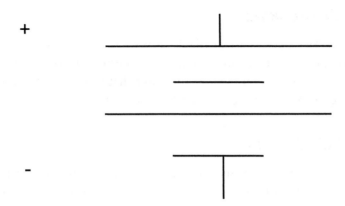

*The polarity of the voltage source's terminals is labeled with positive or negative signs.*

**HIGH VOLTAGE!**

People originally believed that the flow of electrical current moved from the positive to the negative terminal because they thought the positively charged terminal had "excess" electricity. That is called *conventional flow*. We now know that current flows from negative to positive; this is called *electron flow*. Some circuit diagrams depict conventional flow whereas others depict electron flow. All diagrams in this book represent electron flow, but be aware that you may encounter diagrams that depict conventional flow.

## Short Circuits

When current flows through a circuit, it must encounter some resistance because otherwise it can create a short circuit. A short circuit occurs when the current takes an unintentional path or unintended shortcut—it's like a train jumping a rail. The preceding example uses current to power the lamp, providing for a drop in voltage after the current encounters the resistance provided by the load of the lamp. Without the lamp, the voltage doesn't drop along a conductive piece of wire before the current is returned to the positive terminal. Heat is produced along the wire due to friction, and the heat can damage the chemicals in the battery or cause heat damage to the wire along the circuit.

You must be careful to create a circuit that takes into account Ohm's Law. If your circuit has too much resistance built into it, the power it provides to the load may not be adequate to make it work as intended. If you don't provide enough resistance, you might exceed the power that the load can handle. An overloaded circuit can cause excessive heat and the components could explode or cause a fire.

## Fuses and Circuit Breakers

Two types of protective devices are available to prevent a short circuit: *fuses* and *circuit breakers*. Both components create an open or broken circuit if too much current flows in the circuit. A fuse is a single-use component, meaning that it must be replaced if used to create a break in the circuit. A circuit breaker can be reset repeatedly to protect the circuit.

## Serial and Parallel Circuits

Circuits can be arranged in series (called *serial circuits*) or in parallel (called *parallel circuits*) and even in a combination of the two types. A series circuit moves the current through the components sequentially. A parallel circuit allows the current to flow down two or more paths simultaneously to do more than one operation. Some circuits are called series-parallel and consist of portions that are in series and others that are parallel.

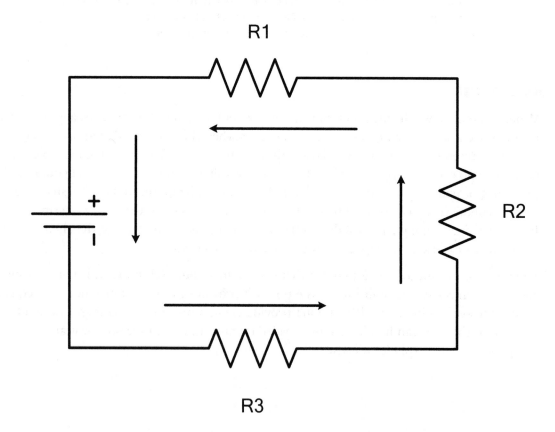

Series

## Parallel

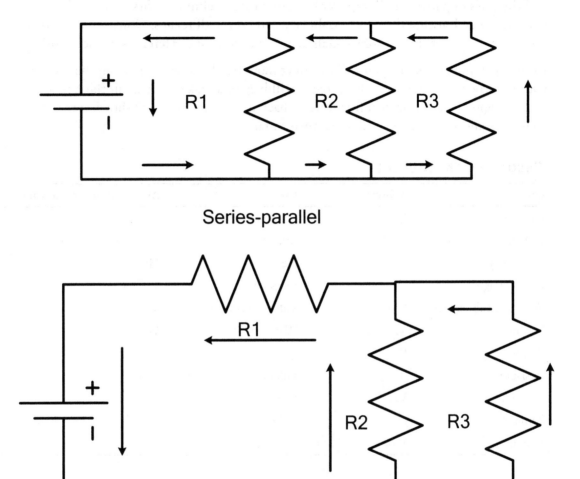

## Series-parallel

*Examples of a series circuit, parallel circuit, and a series-parallel circuit. Resistors are labeled R1, R2, and R3.*

# Learning the Language of Electronics

If you have ever studied a foreign language, you know that the vocabulary, the grammar, and the ways of putting it all together can seem overwhelming at first. Up until this point, you have been learning the basic vocabulary and how it all relates. Don't worry if you can't remember every bit. Even professionals keep handbooks and references at their workbench.

In the upcoming chapters, you will be doing more hands-on work, putting the theory you've learned into practice. This process will help you make sense of all these new terms and concepts. Meanwhile, feel free to refer back to this "cheat sheet" showing the major quantities and abbreviations you've learned so far.

## Electronics Cheat Sheet

| Quantity | Abbreviation | Unit | Abbreviation for the Unit |
|---|---|---|---|
| Charge | $Q$ | Coulomb | C |
| Current | $I$ | Ampere | A |
| Frequency | $f$ | Hertz | Hz |
| Peak voltage | $V_P$ | Volts peak | $V_p$ |
| Peak-to-peak voltage | $V_{PP}$ | Volts peak-to-peak | $V_{pp}$ |
| Period | $T$ | Watt | W |
| Power | $P$ | Ohm | $\Omega$ |
| Resistance | $R$ | Volts rms | $V_{rms}$ |
| rms voltage | $V_{rms}$ | Second | s |
| Time | $t$ | Volt | V |
| Voltage | $V$ or $E$ | | |

## The Least You Need to Know

- AC fluctuates in direction of current flow and voltage. The changes in direction and voltage can be described with a waveform.
- Measuring voltage in AC requires analyzing characteristics of that waveform.
- Current flows in a closed circuit; it is interrupted in an open circuit, which results when there is a break anywhere along the path.
- Circuits are represented in a circuit diagram, which uses symbols to represent all the components in the circuit.
- A short circuit results if there is too much current flowing through a circuit. Fuses and circuit breakers can stop the flow to avoid damage to the components along the circuit.
- A circuit can be series (with sequential flow of current), parallel (the current is diverted onto separate paths to power components simultaneously), or a hybrid of series and parallel portions of a circuit.

# Your Workspace and Tools

Now that you understand the basic theory of how electrons move, it's time to get your equipment in order. You need tools to measure, connect, observe, and protect. Some of these are familiar—screwdrivers, pliers, and other hand tools—while others are specialized. Some of these specialized tools are downright ancient (solder and flux) and others, like the digital multimeter (DMM), you'll find you can't live without.

Where should all those tools go? In a clean, well-lighted space, of course! This part discusses various options for setting up your workspace, whether at school or work or inside your home. You will also learn why good work habits are good safety habits. The effort you put into working deliberately will produce better results.

Because electricity can be very dangerous if handled incorrectly, this part stresses the importance of respecting its power by always using proper safety procedures. And should an accident happen, you'll learn what to do to minimize the damage.

# Tools of the Trade

## In This Chapter

- Getting a handle on the tools you'll need
- Using specialty measuring instruments
- Acquiring essential soldering equipment
- Putting safety first

Whether you're a home hobbyist or planning to pursue a career in electronics, you'll need some tools to get started. This chapter introduces you to the essential tools and instruments plus some specialty items for soldering and safety.

Keep in mind when shopping for tools that it's often a better value to spend more money on a higher-quality product than to try to save a few bucks on a low-quality version that probably won't last as long.

## Essential Hand Tools

Good hand tools will last a lifetime. Spend a little extra to get the best quality instruments you can afford. Here are the tools you'll need to get started:

**A lamp with a magnifying glass.** Good lighting with a magnifying glass is essential for working with small components and circuits.

**Wire cutters.** Don't skimp on these: look for a comfortable and solid grip and high-quality blades that will stay sharp. We suggest buying a variety of sizes so that you always have the right size for the job. A good type to look for is a stand-off shear, as this type is less likely to deform the wire or nearby surfaces and leaves behind a consistent length of wire when cutting the leads (the wire portion). Some wire cutters have a lead catcher, a slot that catches wire trimmings so they don't fall onto your project.

*Wire cutters.*

**Wire strippers.** This tool cuts and removes the insulation on coated wire while keeping the wire itself intact. Look for a good-quality handle and blade. Most users will want a simple small wire stripper and a better-quality automatic wire stripper with holes for multiple *gauges* of wire.

**DEFINITION**

Wire **gauge** is a wire's thickness, and determines the amount of current (amperage) a wire can carry. It is measured at the wire's diameter and doesn't include the insulating covering.

**WATTAGE TO THE WISE**

Copper wire is the most common conductor used in electronics. It performs well in most situations and is relatively inexpensive and easy to work with. It is sold as a solid wire or stranded, with multiple smaller wires twisted together, which makes it more flexible than solid wire. Both types of wire are usually insulated polyvinyl chloride (PVC) or neoprene.

**Screwdrivers.** You'll need an assortment of small screwdrivers with Phillips heads (#00, #0, and #1 are good sizes to have) and flat heads (a few in the range of .9 to 3mm). We recommend that you choose nonmagnetized tools, to avoid damaging specialized computer components that use magnetic storage or creating a current by the interaction with other conductive materials. Eventually you will want to get a small set of specialty screwdrivers that includes hex heads, but to start off, standard Phillips and flat head screwdrivers should be sufficient.

**Pliers.** Get an assortment of small pliers, including needle-nose and long-nose pliers. Again, look for good-quality handles and blades.

*Pliers in different shapes and sharpnesses.*

**Tweezers.** Tweezers enable you to grasp the smallest of electronic parts. Again, eventually you'll probably want different sizes, but start with a longish pair (4–6 inches) with a fine tip.

*A good pair of tweezers with a fine tip.*

**A small rotary tool (Dremel-type).** This handheld power tool is essential for drilling small holes; make sure it has attachments for cutting and sanding.

*A Dremel rotary tool.*
(Photo courtesy of Dremel)

**Drill press stand.** This key enables you to make precise holes.

## Essential Instruments

**Scientific calculator.** It is useful to have a stand-alone calculator you can keep at your bench. It doesn't need to be a graphing calculator, which should help keep costs in line. If you haven't had a lot of experience using a scientific calculator, do take the time to read the manual.

*A scientific calculator.*

**An oscilloscope or logic probe.** As noted in Chapter 3, an oscilloscope instrument displays the actual voltages over time as a two-dimensional waveform. A logic probe is a small, handheld tool that measures these states in a digital current. A serious shop or lab will include a good-quality oscilloscope, as it shows more detail and information. However, such an instrument is quite expensive, ranging upward of $400. If you are just starting in electronics and all you need to test for is a one or a zero in a digital circuit, a logic probe may be a more practical choice for you.

An analog scope shows the real-time information of the circuit signal and displays it on the screen as it is happening. A digital oscilloscope or digital storage oscilloscope samples the circuit over and over again and reconstructs the signal on the screen. This enables you to see an event that may happen only once and you can display the event even after it has happened.

If you do decide to buy an oscilloscope, make sure it has adequate bandwidth for the frequencies you will need to measure. A good rule of thumb is to look for a bandwidth of at least five times the maximum signal you expect to be working with, but as the bandwidth increases to these amounts, the prices are quite high (possibly thousands of dollars). For most hobbyist uses, the minimum you should consider purchasing is a dual-trace, 100 megahertz (Mhz) oscilloscope.

*An oscilloscope.*

 **HIGH VOLTAGE!**

You may find used oscilloscopes for very low prices, but make sure there is some warranty that the scope is in good working order and properly calibrated.

**A digital multimeter (DMM).** This is the one tool you will use the most. A multimeter allows you to measure voltage, current, and resistance (as well as many other readings) in DC and AC circuits with an easy-to-read digital display.

DMMs range in price from under $20 to thousands of dollars. As with most tools and instruments recommended in this chapter, buying a well-reviewed, handheld, midprice multimeter will probably be a better value than buying the cheapest instrument on the market, as this tool is going to be your workhorse.

Maximum input voltage (max volts) is the highest voltage your meter can safely read. If you try to use your meter to read a higher voltage, you will damage it and most likely injure yourself. On the meter we use for the projects in this book, the max volts AC and DC is 600 rms, so we will not take a voltage reading on anything that is higher than 600 V AC or DC.

For all the projects and labs in this book, we use the Extech EX210. We like this model because it also includes an infrared thermometer, which is useful for detecting hot spots and for many other tasks.

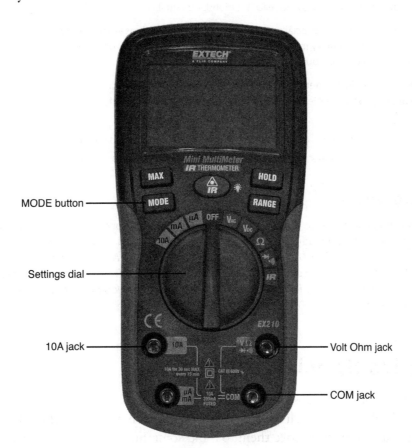

*A digital multimeter.*

Here's what the abbreviations on a DMM stand for:

- **V**AC: volts of AC.
- **V**DC: volts of DC.

- **Ω:** resistance in ohms.
- **A:** whole amps.
- **mA:** milliamps.
- **μA:** microamps (μA).

**HIGH VOLTAGE!**

When using a tool for the first time, or when you get frustrated about a problem that keeps popping up, it really pays to read the friendly manual.

With certain tools, especially expensive ones or in situations where safety is at stake, at least read the rudimentary manuals that are included or go online to locate the resources associated with your new purchase.

Reading an instrument's manual when dealing when electricity can prevent accidents well beyond a simple buzz from a battery. It can save your life.

Using your DMM properly is key to your success with electronics. You can get some hands-on practice by working on the labs at the end of this chapter.

**HIGH VOLTAGE!**

Maximum input amps is the highest current you can test for on your DMM. Most meters have a setting for reading low-value currents in μ or m and a setting for higher current or A. The Extech EX210 DMM , for example, has a jack for μA and mA with a maximum of 200 milliamps (mA) and a jack for a maximum of 10 A. When taking current measurements, always make sure you are using the right settings. You may also need to choose AC or DC; on the EX210, this is done with the MODE button.

# Electronics Specialty Items

**Breadboard.** A breadboard is used in place of a circuit board for testing a circuit. The key value in a breadboard is its reusability. You can place components without having to solder them, and you can easily remove them to build something new. Get as big a breadboard as you can, and make sure you can attach power leads to it.

*A breadboard.*

**Solder.** Solder is a soft metal compound that is melted to join components within a circuit. Do not use plumber's solder; look for solder with a rosin core, 60/40 type (a mixture of 60 percent tin and 40 percent lead). It is sold coiled on a spool.

**Soldering iron.** A soldering iron is the tool used to melt solder to form joints. It has a plastic handle and a metal tip that heats up. You will want a soldering iron that has a variable power control with a maximum wattage of 40. Anything higher can damage the parts or the board. For the projects in this book, we use a Weller WLC 100. This is a good starter iron, but it does have some drawbacks—the iron itself is not grounded and you can control only the power and not the temperature—but for this book and most uses, it is more than up to the task.

*Soldering iron.*

**Solder sucker and solder wick.** A solder sucker is used to suck up excess solder from your projects. A solder wick does the same sort of thing but is more precise. You typically use the sucker first to get the bulk of excess solder off and then use the wick to remove what's left.

**Flux and flux bottle.** Liquid flux is a chemical cleaning agent used to prepare your board for soldering. A flux bottle has a needle that allows for precise application of the flux.

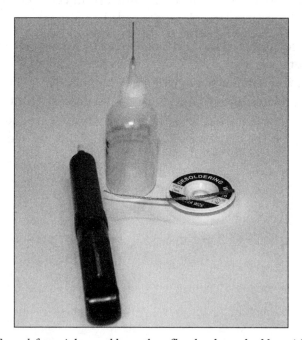

*From left to right, a solder sucker, flux bottle, and solder wick.*

**Heat sink.** A heat sink protects components from heat damage while they are being soldered in place. Get one with an insulated handle, as it will get hot.

*A heat sink with an insulated handle.*

**Circuit board holder.** This is a stand with a set of vice grips for holding a circuit board so you can access both sides of the board.

**Jumper wires.** These are short pieces of wire with the ends stripped for use in breadboard projects. To make your own jumper wires, strip about ½-inch off both ends of varying lengths of different colored wire.

**Clip leads.** These are pieces of wire with alligator clips on each end that are used to make connections on a temporary basis.

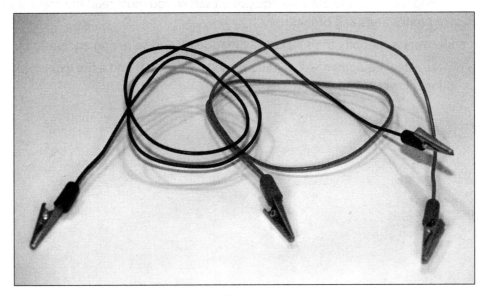

*Clip leads.*

**Variable DC power supply.** This device provides different voltages for project use, enabling you to test and construct different circuits without needing different configurations of batteries and battery holders. You will learn to make one in Chapter 13.

**Function generator.** This is used for injecting different waveforms and frequencies into a circuit. You can precisely control the frequency, which can help when testing and prototyping.

# Essential Safety Items

**Ground (antistatic) strap and ground (antistatic) mat.** These help to prevent electrocution by safely grounding you as you work with electronics. They prevent static discharge.

**Protective eyewear.** You'll need either sturdy glasses or shop glasses.

**Fire extinguisher.** Choose a fire extinguisher rated for Class C fires, which are electrical fires. A $CO_2$ (carbon dioxide) extinguisher is preferable because it doesn't leave a harmful residue.

## The Least You Need to Know

- Screwdrivers, wire cutters and strippers, pliers, and tweezers are all essentials when working on circuits.
- Certain instruments are essential, a scientific calculator and a digital multimeter among them. To operate your equipment safely, you must read the manual to know your tool's limits and proper use.
- Electronics specialty tools include a breadboard and soldering equipment.
- Make safety equipment—including protective eyewear and a fire extinguisher—a priority in your shop.

## Lab 4.1: Taking a DC Voltage Reading

One of the common uses for a DMM is to measure voltage from a power supply or battery. In this lab you practice checking the voltage on an AA battery.

**Materials:**

AA battery

DMM

**Instructions:**

1. On your meter, put the black probe lead in the jack labeled "COM."

2. Put the red probe lead in the jack labeled "V."

3. Move the settings dial to VDC.

4. Touch the black lead to the – or negative side of the battery and touch the red lead to the + or positive side of the battery.

5. Check the reading. It should be around 1.5 V.

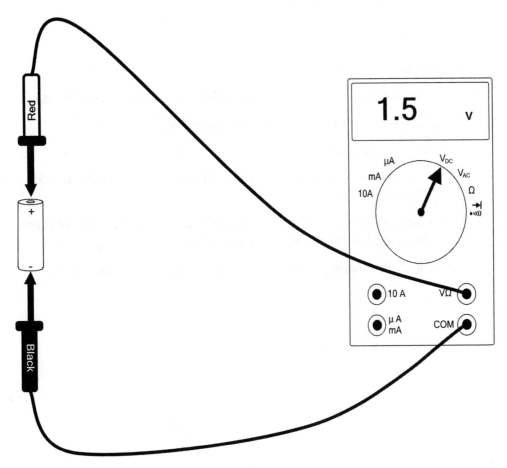

*Reading DC voltage on a DMM.*

## Lab 4.2: Taking an AC Voltage Reading

You can use your DMM to measure voltage from a live wall outlet, which enables you to troubleshoot problems with home wiring.

**Materials:**

Access to a live wall outlet

DMM

**Instructions:**

1. On your meter, put the black probe lead in the jack labeled "COM."

2. Put the red probe lead in the jack labeled "V."

3. Move the settings dial to V$_{AC}$.

    *Very important:* Make sure that your lines are not crossed or touching each other in any way!

4. Place a probe in one slot of a wall outlet and then place the other probe in the other slot. Again, *make sure the probes are not touching each other.*

5. Check the reading. It should register a reading of around 120 V.

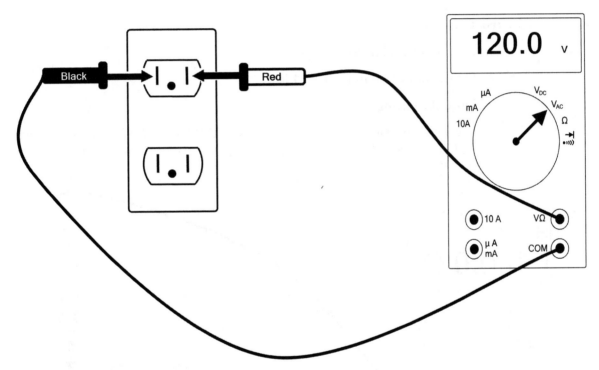

*Reading AC voltage on a DMM.*

## Lab 4.3: Measuring Resistance

A resistor is an electrical component that reduces the voltage in a circuit. In this lab you use your DMM to measure resistance.

**Materials:**

  DMM

  1 100 Ω resistor

**Instructions:**

1. Put the red lead in the V Ω jack.

2. Put the black lead in the COM jack.

3. Move the settings dial to Ω.

4. Use a resistor with a color code of BLACK BROWN BLACK and touch the probes one to each side. The meter should register a reading of 100 Ω.

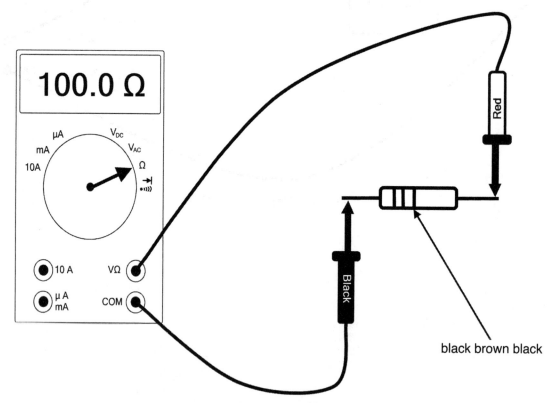

*Using your DMM to measure resistance.*

## Lab 4.4: Measuring Current

You can use a DMM to measure the amount of current flowing in a circuit. Current is measured in amps (A) or in smaller units such as milliamps (mA) or microamps (μA).

**Materials:**

> 9 V battery
>
> Flashlight bulb (lamp)
>
> Two short pieces of wire
>
> Masking tape
>
> DMM

**Instructions:**

1. Tape one end of a piece of wire to the side of the lamp and the other end to the + terminal of the battery.

2. Tape one end of the other piece of wire to the bottom of the lamp, then tape the other end to the red probe.

3. Put the red lead in the 10 A jack.

4. Put the black lead in the COM jack.

5. Move the settings dial to 10 A.

6. Press the Mode button so the display shows DC.

7. Touch the black probe to the – side of the battery. The lamp should come on and you should have a reading of around 0.070 A.

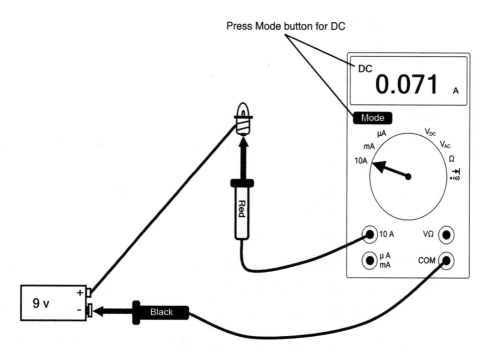

*Using your DMM to measure current.*

# Shop Setup and Safety

## In This Chapter

- Creating a well-lit, organized, and well-equipped shop
- Being prepared and having good work habits
- Avoiding electrical dangers
- Dealing with emergencies

No matter where your electronics shop is located—in your basement or garage, at school, or in a professional shop—you must make safety your highest priority.

If you're working in a professional shop or school, it already has safety rules in place, many dictated by the U.S. Department of Labor's Occupational Safety and Health Administration (OSHA). The guidelines discussed in this book are general in nature and not at all comprehensive. Ask your supervisor for any safety manuals and acquaint yourself with the rules before starting work.

# A Clean, Well-Lit Workshop

The most important safety steps you can make take place before you even begin any projects. As you set up your shop, do so with safety in mind. This involves ensuring you have adequate space, good lighting and ventilation, and well-organized tools.

## Claim Your Space

Whether your workshop is in a stand-alone building or a portion of a garage or basement, make sure you have enough room to store your equipment and work neatly. Ideally, you should be able to close off your workspace, but if you can't, you must be able to somehow prevent unsupervised children or pets from accessing the space. You don't want to have your expensive and potentially dangerous equipment crushed by a stray basketball or used as chew toys.

## Your Workbench

Your workbench should be large enough to hold your instruments and still leave enough work space. The desk or benchtop should be made of a nonconductive material such as hardboard, a synthetic surface that is a good all-purpose material. Any materials labeled as ESD (electrostatic discharge) are good, as they dissipate electrostatic charge. Avoid steel benches, as they conduct electricity.

## Adequate Power

Specialty workbenches usually have built-in power outlets. If you don't want to spend the money on a specialty bench, be sure to situate your unit near adequate power sources. Try to use a circuit of your home that you don't have to share with appliances like minifridges or freezers.

## Lighting

Make sure that you have adequate lighting. Clamp-on lamps are good for spot lighting, but also make sure you have good general overhead lighting. Overhead fluorescents are easy to install and fairly inexpensive solutions for basement and garage setups.

## Ventilation

Good ventilation is essential. Soldering releases lead oxide from the solder and the flux can release irritants. If you don't have access to a window or door to dilute the fumes, you might want to consider purchasing a fume extractor. Small home-use extractors, which cost less than $100, are available from home centers or online sources.

## Storage

Having a good organization system not only makes work easier but also helps keep you and your equipment safe. Tangled cords can become frayed or develop shorts from being bent. Grabbing the wrong tool or component can be disastrous. Improperly stacked equipment can topple over, resulting in injury or damaged equipment.

**HIGH VOLTAGE!**

When working with a tight budget, don't shortchange safety. You can get by without expensive furniture or fancy storage systems, but never skimp on safety equipment or safety procedures.

## Safety Equipment

You must have a good fire extinguisher. Check to make sure it can handle Class C (electrical) fires. A water extinguisher is not going to suppress an electrical fire; in fact, it can cause a much bigger fire.

All fire extinguishers list which fires they are rated for. Dry chemical and carbon dioxide ($CO_2$) extinguishers will both extinguish electrical fires, but $CO_2$ extinguishers will not leave residue on your components. Unfortunately, $CO_2$ extinguishers tend to be more expensive. To determine which is right for you, balance the initial cost against the likelihood of damage.

Protective eyewear is essential. Glasses or shop goggles protect the eyes from wire snips that might go flying when you are cutting or stripping wires.

Grounding straps and a grounding mat (also called antistatic straps and mat) help keep you and your equipment from being damaged by discharging any static electricity that may build up.

# Good Work Habits Are Good Safety Habits

The best way to be safe is to prepare for your project in advance and to work purposefully. Read your equipment manuals, know the limits of your components, and remember your electronic theory. Always double- or triple-check your procedures before adding a component or applying voltage. Good work habits are free but invaluable in terms of safety and getting the job done right.

## Come Ready to Work

Make sure you come to your workbench well rested and wide awake. If you're so tired that your eyes are desperate to stay open, making sense of a complicated circuit diagram will be nearly impossible. Stress can also interfere with your ability to think straight. If you're preoccupied, you're much more likely to make mistakes, such as a slip of your soldering iron. When you come to your workspace, think of it as being like starting up a car. You shouldn't drive drunk, sleepy, or distracted. The same is true for working with electronics. If you are not capable of being completely in control of the situation, don't work with electricity.

## Dress for the Job

Although you don't have to change into overalls or a rubber suit, there are a few wardrobe tips to be aware of when working with electricity. Don't wear jewelry in your electronics workshop. Rings, necklaces, watches, and bracelets are all conductive, and the consequences of adding them to a circuit inadvertently are no laughing matter. Also, keep your hair back and out of your face so you can see clearly.

## Be Neat and Work Deliberately

Electronics is a science of precision, not a freeform enterprise. Work deliberately. Keeping a pad of paper with instructions written out clearly can keep you on task. Put things back where they belong when you are done. Most importantly, be careful about where you set things down. Don't pile components, wires, and instruments on top of each other while you work. You could short out your circuits, start a fire, or damage your equipment.

# Know How Electricity Flows

Current will always flow through the path of least resistance. We have talked so far about constructing circuits, or paths, for current to flow through by design. Remember, though, that electrical current is a natural phenomenon, and charged particles are moving through everything that surrounds us and in your body itself.

If an electric line is broken during a storm and the wires fall to the ground, the current will flow through the ground because a circuit will form between the line and the ground. If you are nearby, you will be a ready part of the circuit because you provide less resistance than the relatively nonconductive earth. Unless you have a good insulator in your rubber soled shoes keeping you out of the circuit, the voltage that flows through the circuit is likely to cause significant injury and possibly death.

You don't want to become inadvertently part of a circuit. If you use your body to connect two terminals of a car battery, you are forming a circuit; it is unlikely that you will ever want to try that experiment again.

**HIGH VOLTAGE!**

Your body is an electrical system that relies on its own circuits to contract and relax our muscles. Your muscles control the pulmonary system by the action of the diaphragm contracting and expanding your lungs; they control the circulatory system by the pumping of your heart. The nervous system is also an electrical grid sending signals from throughout your body to your brain for it to analyze and react to.

Imagine this sensitive system being affected by unwanted current. A large shock can disrupt your muscle function, causing muscle spasms or the locking up of your muscles (this is called *tetanus*). It can also disrupt the rhythms of your breathing or heart rate. The medical field has used controlled shocks to our systems in medical situations to reset or restart our heart rhythms with a large shock of a defibrillator. Law enforcement can use a Taser shock to immobilize an out-of-control suspect.

In addition to keeping in mind the effects of shock from electrical current, remember that your skin and internal organs provide resistance. When a current encounters resistance, it produces heat. One of the most obvious results of electric shock is burning. From a small burn on the skin surface to significant damage to organs, the heat effect of electrical shock is quite dangerous and produces a lot of pain.

# First Aid for Electrical Shock

If you witness someone suffering from electrical shock severe enough to cause unconsciousness, cardiac arrest, seizures, burns, palpitations, muscle pains, or contractions, you should immediately call 911. It is important that the person get professional help right away to prevent further injury or even death.

You can provide the following assistance while you wait for emergency personnel:

- First, always look before touching. The person might still be connected to the circuit. Disconnect the electricity source if possible. Throw the circuit breaker or use a nonconductive material such as wood or cardboard either to unplug the electrical equipment or move the person away from the electrical source.

- If the victim is separated from the electrical source, provide cardiopulmonary resuscitation (CPR) if you don't notice any pulse or breathing.

- If possible, and if you can do it safely without risking electrocution, lay the person down and elevate his or her legs to reduce the risk of shock.

If you experience a small shock that doesn't leave a burn, no treatment is usually necessary. But if there is burning, especially if the current traveled through your body, you should seek care at the emergency room. A shock that leaves just a small visible burn may do significant internal damage. Large shocks can damage internal organs and should be examined by a doctor.

Water decreases the resistance of your skin. Never handle electronics with wet hands. Now, we all know not to handle electric appliances in the shower or bath, but think of other ways your hands can get wet. Holding a glass that has water condensed on it will give you wet hands. Sweaty hands decrease resistance even more, as the salty water of sweat is more conductive than plain water.

## The Least You Need to Know

- A good workspace requires good lighting, ventilation, safety equipment, and organization.

- Good work habits are good safety habits. Don't come to your projects impaired. Be prepared and work deliberately.

- Current will travel the path of least resistance. Make sure that you don't accidentally use your body to complete a circuit.

- Electrical shock is serious. Call 911 immediately in cases of severe shock and seek medical attention for any shock that results in a visible burn.

# Electronic Components

This part gets into some of the workhorse components of electronics: switches and resistors. Every circuit will rely on them so you need to understand the various types and how they are represented on circuit board diagrams.

And once current is flowing to the components of the circuit, you need to be able to fine-tune the flow. Capacitors and diodes make sure that the flow is just so, avoiding spikes and drops. Capacitors can also store power to provide quick bursts when needed.

Since their invention in the 1950s, transistors have been as important to electronics as the portable scientific calculators they made possible. Transistors enable you to control the amount of current that flows through a circuit, diverting it or even reversing it. Their invention and use made possible modern communication and all of the other great electronics advances of the twentieth century. Transistors ability to manipulate the flow of electricity has enabled electronics to create logic gates that can be used to make calculations, which is the very basis of modern computers.

# Switches

## In This Chapter

- Using a switch to open and close a circuit
- Making sense of poles and throws
- Identifying different types of mechanical switches
- Getting acquainted with relays

A switch is the most basic component in a circuit. It has the basic function of interrupting the current flow and creating an open (or broken) circuit. You are already familiar with some switches, such as light switches and on/off switches on any of your electric or electronic devices. Other switches include keys on a keyboard and buttons on a car stereo.

Switches always control simple open/close operations; they are not used for fine-tuning varying levels of current in a circuit. Switches also are essential to the many subcircuits in more complicated devices.

# Switch Symbols

On a circuit board, switches are represented as being either open or closed, as shown in the following illustration.

*The symbol for an open switch in a circuit diagram.*

**WATTAGE TO THE WISE**

The open/closed states of a switch can be represented by the *binary* digits 1 and 0. This simple concept is essential to the creation of digital electronics. See Appendix D for more information about binary numbers.

# Mechanical Switches

One of the most basic types of switches is a mechanical switch. The conductive parts of mechanical switches are called *contacts*. When in the closed state, the contacts allow current to flow through the circuit. A mechanical force called an *actuator* brings the contacts together or apart. The force required to move the actuator is called *actuation force*.

There are a wide variety of mechanical switches available, and each of them is represented by a specific symbol on circuit boards. The following sections describe some of the more common mechanical switches.

## Poles and Throws

Switches can control either a single circuit or multiple circuits. A switch is described by the number of *poles* and *throws* it can make—for example, single-pole, double-throw (SPDT) or double-pole, double-throw (DPDT).

**DEFINITION**

In a switch, a **pole** is each contact that completes a circuit. For instance, a typical household light switch controls one circuit. It then is said to have a single pole. If you flip a switch that turns on a light and a fan simultaneously, it has two poles.

A **throw** is one of the several positions that will allow a switch to make pole connections. Both of the switches just described have only one throw. The switch that turns on the light is a single-pole, single-throw (SPST). The switch that controls two circuits—one to the light and one to the fan—has one throw, so it is a double-pole, single-throw (DPST).

Think of poles as the number of things that can be controlled, and the throws as the actions needed to make the connections. So if you have a simple on/off switch that controls three circuits, it is a triple-pole, single-throw (TPST) switch.

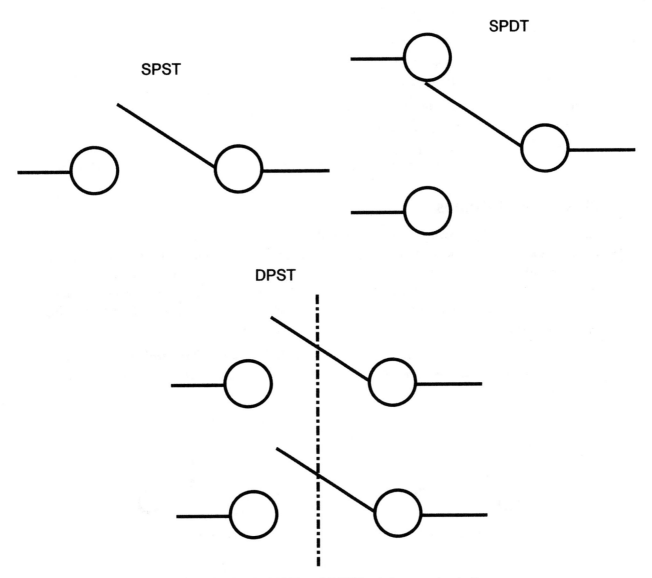

*The symbols for SPST, SPDT, and DPST switches on a circuit diagram.*

## Some Common Switch Abbreviations

| Abbreviation | Switch Type |
| --- | --- |
| N.C. | Normally closed circuit |
| N.O. | Normally open circuit |
| SPST | Single-pole, single-throw |
| SPDT | Single-pole, double-throw |
| DPST | Double-pole, single-throw |
| DPDT | Double-pole, double-throw |
| MBB | Make before break |
| BBM | Break before make |

## Push-Button Switches

In a push-button type switch, the mechanical force either makes the connection between the contacts or breaks it. Push-button switches are classified as either normally open (n.o.) or normally closed (n.c.). An n.o. push-button is a push-to-make switch, meaning that when you push the button it makes the contact, closing the circuit. An n.c. push-button is a push-to-break switch, meaning that the contact is made until you push the button, making a break in the circuit.

A similar way to classify switches is as either *break before make* (BBM), which is the same as an n.o. switch, or a *make before break* (MBB) which is the same as an n.c. push-button switch.

Normally Open

Normally Closed

*The symbols for push-button switches on a circuit diagram.*

## Knife Switches

A knife switch is an SPST switch. It is constructed of a strip of conductive material with an insulated handle on a hinge. When the hinge is closed, the circuit is closed as well. When it is opened, the circuit opens and the flow of current is interrupted.

This basic switch doesn't have many applications in home electronics, but think of a simple switch in a movie laboratory, like the one Dr. Frankenstein used to create his monster. Large high-voltage switches, like the main power switch in industrial or commercial settings, are examples of modern uses of knife switches.

## Bi-Metal Switches

A bi-metal switch is made of two different metals with different sensitivities to heat. The metals are coiled together in layers and can rotate as heat rises due to the metal's expansion when heated and its contraction in response to the metal's cooling. When the rotating coil reaches a certain contact point, it activates a switch. Bi-metal switches are very important in heat regulation applications such as in an oven or in an automobile's temperature control.

**HIGH VOLTAGE!**

In high-power situations, there is a danger of arcing (pronounced *ARK-ing*) as contacts come apart. The insulating air between the contacts can become ionized and the gas/current mixture forms an *electric arc*. There are several ways to prevent this potentially dangerous and destructive event. One technique is to enclose the switch in an insulating gas mixture such as sulfur hexafluoride; another is to use a magnetic blowout that redirects any arc.

## Mercury Switches

A mercury switch is operated by the gravitational effect on a ball of mercury in a vacuum tube. The mercury will always go to the lowest part of the tube. If the tube is rolled to the side, the mercury will change where it makes contact. The advantages of this type of switch are that it will not produce a spark in an environment where flammable gases may be present and it doesn't corrode from repeated metal-on-metal contact.

In the past, thermostats often used mercury switches in combination with a bi-metal switch because they didn't degrade over thousands of uses. Because mercury is toxic, these types of switches are no longer used in many consumer applications.

## Other Mechanical Switches

You'll encounter many other types of switches if you work with electronics for very long. The type of switch used is often a matter of design, availability, or the level of security involved; often it has nothing to do with electronic function. A slide switch can have one or more poles, with each stop on the slide making or breaking the connection.

A toggle switch is operated by a lever or rocking mechanism. Think of the little toggle switch on a guitar amplifier.

A rotary switch has multiple pins that when rotated can make a connection. An example is the knob of a three-speed fan or the cycle selector on a basic washer or dryer.

A rocker switch is the type of switch that you may see on your surge protector; a push on the raised portion of the switch will activate the other position.

A keylock switch is activated with the action of a key in the lock; think of your car's ignition as one example.

## DIP Switches

One type of switch that you won't see as a visible part of many consumer applications is a dual-inline-package (DIP) switch. It is a series of switches in one unit that can be set individually. In the past, this type of switch was often used on printed circuit boards to allow for manual adjustment of the settings. There are still many industrial uses for this type of switch. DIP switches can be any of a number of different mechanical switch types, but the term *DIP switch* refers to this type of setup.

**WATTAGE TO THE WISE**

When a mechanical switch is actuated and the contacts meet, they may create a bit of noise (an unwanted signal that interferes or obscures the electrical signal you intend to transmit) by the physical properties of the two contacts coming together. This is called *contact bounce.* There are some methods to filter this noise in more sophisticated or sensitive electronics.

# Electromagnetic Switches or Relays

A switch that uses an electromagnetic field to control the opening and closing of a circuit is called a *relay.* One modern use of a relay is in the hinge of a laptop, to put the computer into hibernation mode as the cover is closed; another is a proximity switch for a burglar alarm, which detects when the contacts are separated by the opening of the window or door.

One type of relay is the reed switch, which was invented in 1936 at Bell Labs. It is made by applying a magnetic field onto two thin strips of ferrous metal in an enclosed tube. When the field is present, the circuit remains closed. If the field is interrupted, the circuit is broken.

## The Least You Need to Know

- A switch is an electronic component that can open or close a circuit. Contacts are the conductive materials that, when brought together by the switch, complete (close) the circuit.

- There are several types of mechanical switches, including the push-button, knife, bi-metal, and mercury switch.

- Switches are classified by the number of poles and throws they have. A pole is the position where the circuit is closed; a throw is one of the several positions that will allow a switch to make pole connections.

- Electromagnetic switches are called relays. The electromagnetic field brings the contacts together or pulls them apart.

# Lab 6.1: SPDT Switch

In this lab you'll construct a simple circuit with a switch and lamps to help you understand the parts and function of a switch.

**Materials:**

> 9 V battery
>
> 1 SPDT switch
>
> 1 breadboard
>
> 2 small lamps
>
> Jumper wire

**Instructions:**

To do this lab you need to know how to use a breadboard, an essential piece of electronics equipment that provides a convenient way to make nonpermanent, solderless connections between components.

A breadboard has a top and bottom portion called a power rail. Each power rail has two rows of connection plugs. Whether you are using a battery or power supply, you connect your power in these rows. Keep the positive lead connection on the top left and the negative lead (for a battery) or GROUND (GND) lead (for a power supply) connection on the bottom left.

Each individual hole in the breadboard is called a cell, and the various cells can be connected by plugging jumper wires into them.

The two larger rows between the power rails are the work areas where you will make your circuit connections when you need more than just a power connection.

Now that you know how to use a breadboard, let's get started:

1. Connect the jumpers and battery to your breadboard as shown in the diagram.

2. Flip the switch to the right. One lamp should be on.

3. Flip the switch to the left. The other lamp should be on.

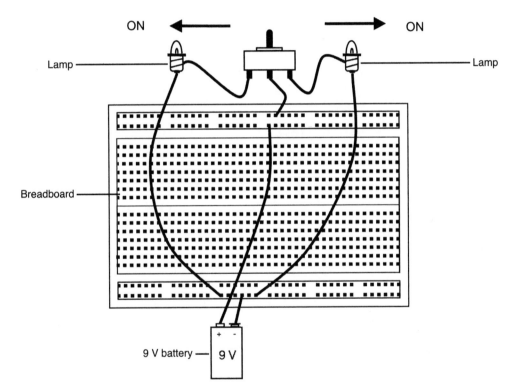

*Constructing a single-pole, double-throw switch.*

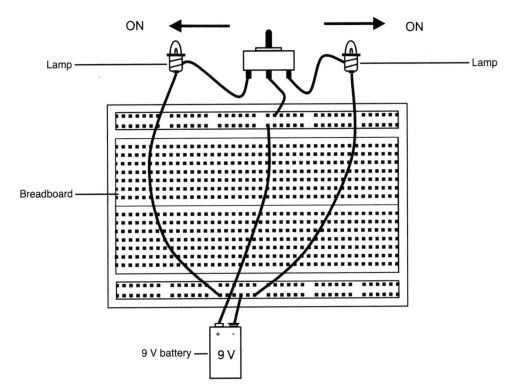

ON

ON

Lamp

Lamp

Breadboard

9 V battery

9 V

*Constructing a single-pole, double-throw switch.*

# Resistors

## In This Chapter

- Using resistors to control current
- Making sense of the many types of resistors
- Reading a resistor's color code
- Using Ohm's Law to calculate resistance

Resistors are electronic components that increase resistance. They work against the flow, reducing the amount of current in a portion of a circuit. When you add a resistor into a segment of a circuit, the current that flows out of the resistor is less than the current that entered it.

Resistors dissipate the current through heat and are classified by the amount of energy they can dissipate. Most resistors are made of ceramic or other materials that radiate heat at a predictable rate.

## The Mighty Resistor

By managing the amount of voltage pushing the current through a circuit, resistors protect components that can only handle certain voltages. Resistors also can serve as voltage dividers by partitioning voltage into smaller values to perform precision tasks. Because resistors make it possible to fine-tune the current flow, they are major electronic components. The following figure shows the symbol for a resistor on a circuit board diagram.

*The symbols for a resistor in a circuit diagram.*

# Fixed-Value Resistors

A fixed-value resistor, sometimes simply called a fixed resistor, is a component designed to provide a stable resistance. There are several types of fixed resistors, including carbon-film, metal-film, ceramic, and wire-wound. Each relies on layers of resistant materials and is categorized by size, level of resistance, and *tolerance*.

**DEFINITION**

**Tolerance,** in electronics, is the percentage in possible variation from the stated value.

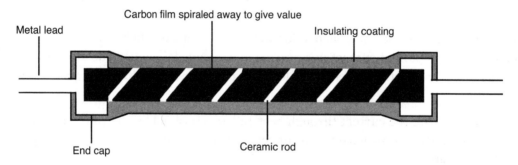

*A carbon-film resistor.*

In a carbon-film resistor, a small ceramic rod surrounding a metal-wire lead is covered in carbon film. The carbon film is removed in spirals to achieve the particular value of resistance required. The whole mechanism is covered in an insulated material, and a color-coded band labels the resistor.

# Resistor Color Codes and Power Ratings

In the 1920s, the Radio Manufacturers Association standardized the Resistor Color Code (see the following table). It consists of three or four colored stripes that represent the value of the resistance provided. The first two stripes are the significant digits (the digits to the left of the decimal point), whereas the third digit is the exponent value (the powers of ten). The formula for calculating the resistance is as follows: $AB \times 10^x$, where A represents the first significant digit, B represents the second significant digit, and $10^x$ is the exponent of 10 value.

The fourth stripe of the resistor code, if present, represents the stated tolerance of the resistor. If there is no fourth stripe, the tolerance is plus or minus (±) 20 percent. This means that the resistor may provide up to 20 percent more or 20 percent less resistance; so with a resistor with a value of 100 Ω, its true range is anywhere from 80 Ω to 120 Ω.

The following table shows the resistor colors and their assigned values:

## Resistor Color Code

| Color | 1st Band | 2nd Band | 3rd Band (Multiplier) | 4th Band (Tolerance—%) |
|---|---|---|---|---|
| Black | 0 | 0 | $\times 10^0$ | N/A |
| Brown | 1 | 1 | $\times 10^1$ | ± 1 |
| Red | 2 | 2 | $\times 10^2$ | ± 2 |
| Orange | 3 | 3 | $\times 10^3$ | N/A |
| Yellow | 4 | 4 | $\times 10^4$ | N/A |
| Green | 5 | 5 | $\times 10^5$ | ± 0.5 |
| Blue | 6 | 6 | $\times 10^6$ | ± 0.25 |
| Violet | 7 | 7 | $\times 10^7$ | ± 0.1 |
| Gray | 8 | 8 | $\times 10^8$ | ± 0.05 |
| White | 9 | 9 | $\times 10^9$ | N/A |
| Gold | | | $\times 10^{-1}$ | ± 5 |
| Silver | | | $\times 10^{-2}$ | ± 10 |
| None | | | | ± 20 |

## Reading the Code

Let's determine the resistance of a resistor with a blue stripe, a yellow stripe, and an orange stripe. First, refer to the Resistor Color Code to determine the values for each stripe. According to the Resistor Color Code, the first stripe, blue, has a value of 6. The second stripe, yellow, has a value of 4. The third stripe has a value of $10^3$, or 10,000 (see Appendix D for information on the power of 10). To calculate the resistance level, you multiply 64 times 10,000, giving you a value of 64,000 Ω.

Let's consider a resistor that is marked, from left to right, yellow, violet, and red. The first band is yellow, so the first digit is 4. The second band is violet, which represents 7. The third stripe is red, which represents $10^2$. So we would have a resistor value of $47 \times 10^2$ or 4,700. As there is no fourth band, the tolerance is ± 20 percent.

**WATTAGE TO THE WISE**

Only the tolerance band can be gold or silver, so if that is the first band you see, flip the resistor over and start reading the code from the other end.

The color code is an imperfect system. Overheating of the resistor can make the colors hard to distinguish, and colorblind individuals can't rely on the system. To account for these issues, some resistors are also stamped with a numeric value as well.

## Power Ratings

Resistors are classified not only by their resistance but also by their power rating, which represents the highest amount of power a resistor can withstand. As currents travel through the resistor, heat is released. If you use a resistor with too low of a power rating, it can fail or cause damage.

The power rating is given in watts. Because watts are a unit of power and not current or voltage, to determine the appropriate power rated resistor for your projects you need to calculate the wattage of your circuit. To solve for watts, you multiply the current in amps by the voltage in volts: $P = I \times R$.

Carbon-film and metal-film resistors are usually available in a range from $\frac{1}{8}$ to 2 W. Usually, the larger the physical size of the resistor, the higher the power rating. Wire-wound resistors are used where higher wattage ratings are needed, as they can have a power rating range from 1 W to 10 W.

# Surface Mount Resistors

Traditional resistors described earlier in this chapter have two metal leads with a ceramic core. Modern electronics generally use surface mount technology, mounted via soldering to the surface of a board. Surface mount resistors do not use the resistor color code. Instead, a three- or four-digit code is printed on the resistor. In a three-digit code, the first digit represents the first significant digit, the second, the second significant digit, and the third is the exponent of 10 value. So if it is marked 201, the resistance value is $20 \times 10^1$, or 200 $\Omega$. Some precision applications require more precise resistors; in that case, you may see a four-digit code. The first three digits represent the significant digits and the fourth digit is the exponent of 10 value.

# Single In Line Resistors

*Single in line* (SIL) resistors are a linked series of resistors in one combined package, sometimes called an SIL resistor network. They are used in many home electronics; due to their compact size they require less solders than multiple individual resistors. They are primarily used in surface mounted circuits.

*The SIL resistor pictured here has nine resistors, each delivering 47 $\Omega$. The code reads 470, which equals $47 \times 10^0$, or 47.*

# Variable Resistors

Some resistors do not carry a fixed resistance value and are called variable resistors. Types of variable resistors include thermistors (variable due to temperature) and photoresistors (variable due to levels of light), also known as photocells. Another variable resistor is the varistor (variable + resistor), which serves mainly to protect sensitive components from short spikes of excessive current. When triggered, it shunts the current away from the more current-sensitive components.

One of the most commonly used variable resistors is the potentiometer (often simply referred to as *pot*), which is familiar to anyone who has used a volume knob on an analog stereo or television. Instead of the simple two terminals (or leads) of fixed-value resistors, a potentiometer has three terminals. The third terminal in a variable resistor can be moved to different tapping points to yield varying levels of resistance. The terminal that moves is usually in the center of the three-terminal arrangement and is called the *wiper*.

A pot is usually controlled by a knob that moves the wiper to the different tapping points. Older televisions used tone knobs for the control of brightness, and car and home radios were once controlled by pots. Any analog signal can be controlled with a pot.

Resistors have different symbols depending on whether they are fixed-value or variable resistors. Here are the symbols you would find in a circuit diagram for several of the most common types:

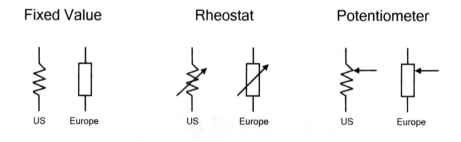

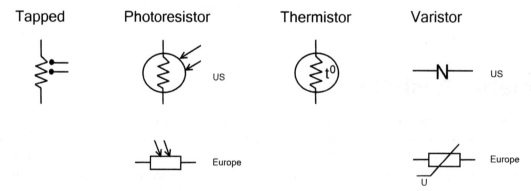

*Symbols for various types of resistors in a circuit diagram.*

# Kirchhoff's Laws

To calculate the proper resistance to use in a particular part of a circuit, it is necessary to look at the energy in that circuit. You already know that voltage pushes electrons, creating an electrical current. To understand what happens when you introduce resistance to that current, let's look at another foundational concept of electronics: Kirchhoff's laws.

Gustav Kirchhoff described two basic laws of electronics that follow from the concept of *conservation of energy.* The first relates to current in a *junction:* the amount of current that enters a junction is the same as the current that exits that junction.

**DEFINITION**

**Conservation of energy** is the principal that energy is never lost or gained without the application of an outside force. Along with conservation of mass, it is one of the fundamental laws of physics.

A **junction** is each particular meeting of wire and component along a circuit.

The second of Kirchhoff's laws states that the sum of all voltages in a closed circuit is zero. The current pushed through a circuit is met with the resistance of the different components, each of which generate negative voltage, also called a voltage drop. All of the positive push of the voltage is matched by the combined negative push of the resistance. In fact, resistance is sometimes expressed as negative values of voltage. Resistors are not the only components providing resistance; the load (lamp, motor, diode, etc.) and wire of the circuit itself provides some resistance.

So using Kirchhoff's laws we know that all of the voltage in a circuit (the power supply or battery) must equal the total resistance in a circuit.

# Calculating Resistance

In Chapter 2, you learned about Ohm's Law, which states that current (I) between two points is directly proportional to the voltage (V) and inversely proportional to the resistance (R). You learned that as an equation, it is written $I = V/R$, and that if you have any two of the variables, you can solve for the other: $V = R \times I$ or $R = V/I$.

Based on this law, as you construct a circuit you have three variables to consider: voltage, current, and resistance. In practice, you will encounter components that have specified current and voltage levels for their operation, and you will need to use Ohm's Law to calculate the amount of resistance that must be provided between the power supply and the component. In other words, you will be solving for R. The equation for solving for R is R = V/I, which means to find R, you need to divide the voltage by the current.

Here is a basic example. You have a 5 V power supply that provides 500 mA of current. If you want to construct a circuit that uses a component that requires no more than 20 mA of current, you can solve for the resistance needed by dividing 5 V by .02 A. The desired resistance would therefore be 250 Ω.

Notice that the values for the current in the preceding example were expressed in milliamps (mA) instead of amps. As you work on electronics, not all values will be expressed as amps (A), volts (V), or ohms (Ω). You are regularly going to encounter values such as mA or micro-amps (μA). To do any calculations, you need to make sure that you convert your variables into like terms. So if you are dealing with 8 V and 500 mA, you need to convert your mA to A by either expressing the number as .5 A or $500 \times 10^{-3}$ A.

To help you with these conversions, consult the following table for a list of basic metric prefixes and their values:

## Metric Prefixes and Their Values

| Number | Exponent value | Prefix | Symbol |
|---|---|---|---|
| ×1,000,000,000,000,000 | $10^{\times 15}$ | femto | f |
| ×1,000,000,000,000 | $10^{\times 12}$ | pico | p |
| ×1,000,000,000 | $10^{\times 9}$ | nano | N |
| ×1,000,000 | $10^{\times 6}$ | micro | μ |
| ×1,000 | $10^{\times 3}$ | milli | m |
| 10 | $10^{0}$ | N/A | N/A |
| 1,000 | $10^{3}$ | kilo | K |
| 1,000,000 | $10^{6}$ | mega | M |
| 1,000,000,000 | $10^{9}$ | giga | G |
| 1,000,000,000,000 | $10^{12}$ | tera | T |

Once you know the level of resistance needed in the circuit, you can use the right resistors to achieve the total resistance required. You need to consider the resistance of the components and then combine different values of resistors to achieve the overall resistance required. Remember that resistors have different tolerances, so there may be a variation of up to 20 percent. Most circuits are designed with those variations in mind, but some will require more precise values.

## Resistors in Series Circuits

Resistors can be arranged in series, which means they can be placed one after another in a circuit. To determine the total resistance provided by resistors in series, you simply add together their values $R_T = R_1 + R_2$, where $R_T$ represents total resistance and $R_1$ and $R_2$ represent the first and the second resistors in series combined.

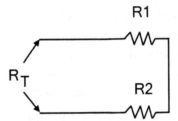

*The symbol for resistors in series on a circuit diagram.*

**WATTAGE TO THE WISE**

When working through these labs, it is important to follow the instructions in conjunction with the images or circuit diagram accompanying them. In most electronics projects, no formal written instructions are provided, just a schematic, or enhanced circuit diagram, and some notes. Learning to interpret the schematic is a critical step in becoming adept in the field of electronics.

## Resistors in Parallel Circuits

If resistors are arranged in a parallel circuit, the math determining their total resistance is a little more complicated. To determine the total resistance provided by three resistors that are in parallel, use the following equation: $1/R_{equivalent} = 1/R_1 + 1/R_2 + 1/R_3$. If you have multiple resistors, the equation would just continue on, with further resistor values being entered in a like fashion ($1/R_x$).

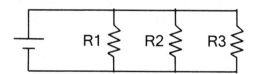

*The symbol for resistors in parallel on a circuit diagram.*

A good rule of thumb is that in a parallel circuit, the equivalent resistance will always be lower than the lowest value resistor.

## Voltage Division Circuits

A voltage division circuit is used in situations when you need to supply varying output voltages or to interpret the different voltages produced by the circuit to provide information. The first use is fairly straightforward: you have a power supply that provides one voltage, and you need to reduce it along the circuit to produce lower voltages to different components. The second use is a bit more complicated but incredibly useful.

As mentioned previously, variable resistors provide different levels of resistance depending on how the resistor was set. A potentiometer might be used to increase or decrease the resistance to lower or raise the volume on a stereo. But the resistance that is detected in variable resistors like the photoresistor or photocell can provide information about an outside value—the amount of light detected, for example—that affects the output voltage. You can then use the outputted voltage to feed information into the circuit about how bright of a light is detected. This means that a photoresistor can act as a light sensor (see Chapter 21 for more on sensors).

Let's focus on the first use of a voltage divider circuit. Suppose you want to be able to produce a certain outputted voltage, known as voltage out. To determine the voltage out in a simple voltage division circuit, the equation is $V_{out} = V_{in} (R_2 / R_1 + R_2)$.

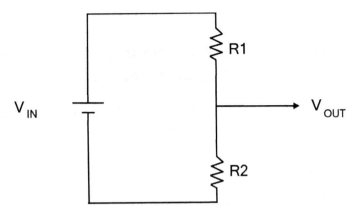

*The symbol for a voltage division circuit on a circuit diagram.*

## The Least You Need to Know

- Resistors are components that increase the resistance in a circuit.
- There are fixed-value resistors and variable resistors. Fixed-value resistors are marked with a Resistor Color Code indicating their value.
- One type of variable resistor is the potentiometer. An example of this is a volume knob on a stereo.
- You can use equations derived from Ohm's Law to determine the total resistance of resistors in a circuit.

# Lab 7.1: Using Ohm's & Kirchhoff's Laws to Determine the Proper Resistor

In this lab you build a simple circuit with a red LED (light-emitting diode) as a load. You want to provide a voltage of approximately 2 V and 20 mA (.020 A) of current to power the red LED, and you need to calculate the value of resistor you need.

**Materials:**

      Breadboard

      1 9 V battery

      1 red light-emitting diode (LED)

      1 390 Ω resistor (orange white brown)

      Jumper wires

The supply voltage of our battery ($V_S$) is 9 V. The diode requires 2 V, so our total voltage is 7 V (9 V – 2V = 7 V). It requires 20 mA of current. When doing calculations in Ohm's Law, you need to use amps as the standard unit, so you should use .020 A (the equivalent of 20 mA) in your calculations. R = 7 ÷ .020, which yields a required resistance of 350 Ω. The closest value resistor available is 390 Ω.

**Instructions:**

1. Place the components on the breadboard as shown in the diagram. Note the placement of the red LED.

2. Attach the jumper wires to the battery as shown. The lead attached to the – side of the battery *MUST* be attached to the flat side of the LED.

You should see that by using Ohm's Law you have created a circuit that provides adequate voltage and current without overwhelming the tolerances of the load. The LED lights up and there is no excessive heat or damage to the LED.

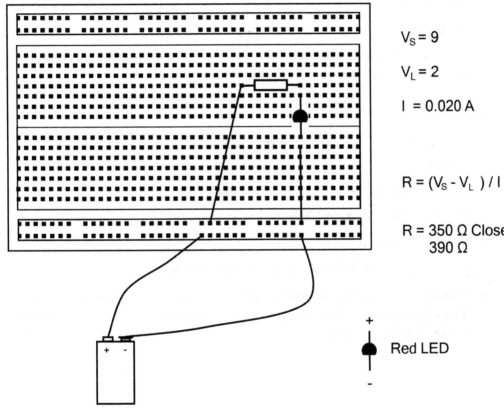

$V_S = 9$

$V_L = 2$

$I = 0.020$ A

$R = (V_S - V_L) / I$

$R = 350\ \Omega$ Closest value is
390 $\Omega$

Red LED

# Lab 7.2: Resistors in a Series Circuit

In this lab, you will use your DMM to see how resistors affect the measured voltage when you arrange three in a series.

**Materials:**

Digital multimeter (DMM)

Breadboard

Jumper wire

Three resistors with values of 470 Ω (yellow violet brown), 100 Ω (brown black brown), and 270 Ω (red violet brown)

**Instructions:**

1. Place the resistors in a series on the breadboard exactly as shown in the diagram.

2. Place the jumper wire on the breadboard as shown.

3. Put the black lead in the COM jack of your meter.

4. Put the red lead in the jack labeled VΩ.

5. Move the selector dial to the Ω position on the meter.

6. Touch the two probes one to each jumper wire coming off the breadboard. The value you get will vary depending on the tolerance of your resistors but it should be close to 840 Ω.

Chapter 7: Resistors    93

R1= 470 Ω

R2= 100 Ω

R3= 270 Ω

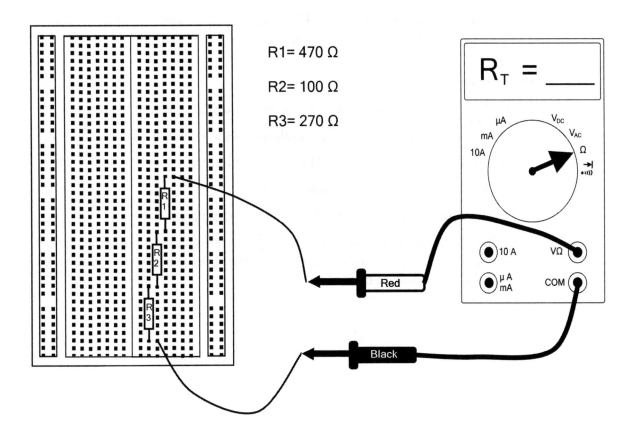

## Lab 7.3: Resistors in a Parallel Circuit

**Materials:**

DMM

Breadboard

Jumper wire

Three resistors with values of 470 Ω (yellow violet brown), 100 Ω (brown black brown), and 270 Ω (red violet brown)

**Instructions:**

1. Place the resistors in parallel on the breadboard as shown in the diagram.

2. Place the jumper wire on the breadboard as shown in the diagram.

3. Put the black lead in the COM jack of your meter.

4. Put the red lead in the jack labeled VΩ.

5. Move the selector dial to the Ω position on the meter.

6. Touch the two probes one to each jumper wire coming off the breadboard. The value you get will vary depending on the tolerance of your resistors, but it should be close to 63 Ω.

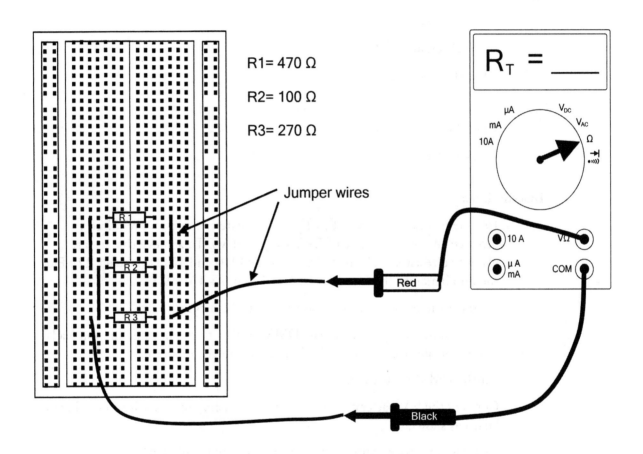

R1= 470 Ω

R2= 100 Ω

R3= 270 Ω

Jumper wires

## Lab 7.4: Voltage Division Using Fixed-Value Resistors

In this lab you will build a voltage divider circuit using fixed-value resistors. You will see if the calculations you do to determine the voltage out matches what your DMM measures.

**Materials:**

> 1 470 Ω resistor
>
> 1 100 Ω resistor
>
> 1 breadboard
>
> 1 9 V battery
>
> 1 DMM
>
> Jumper wire

**Instructions:**

1. Use the formula $V_{out} = V_{in} \left( \frac{R_2}{R_1 + R_2} \right)$ to calculate the resistors required to produce an output voltage of 7.4 V ($V_{out}$ = 7.4 V). To achieve that we are using two resistors valued at 100 Ω (labeled R1) and 470 Ω (labeled R2). Plug the numbers into the formula: 7.4 V = 470/100 + 470 × 9 V.

2. Connect the resistors in series as shown in the diagram.

3. Using jumper wires, connect the DMM probes to the bottom and top leads of the bottom resistor, labeled R2, as shown in the diagram.

4. Set the DMM to measure DC voltage.

5. Connect the 9 V battery's + and – terminals using jumper wire to the breadboard as shown.

6. Using jumper wire connect the two sides of the breadboard to the series resistors as shown.

7. Your DMM should show a value for this outputted voltage $V_{out}$ that is close to what we calculated above: 7.4 V.

R1 = 100 Ω
R2 = 470 Ω

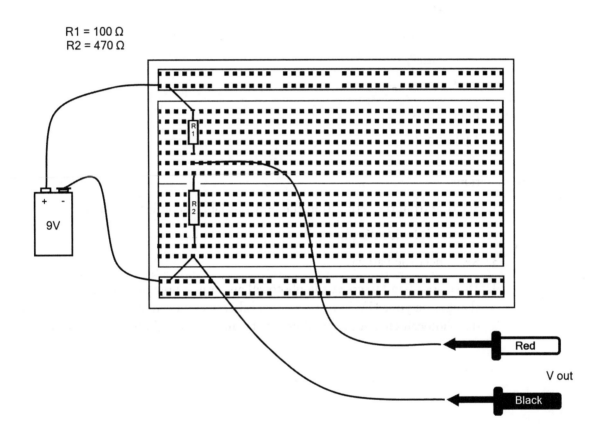

## Lab 7.5: Voltage Division Using a Variable Resistor

In this lab you will build a voltage divider circuit using a variable resistor called a photoresistor or photocell. You will see the reaction of the photocell to varying light conditions and how it affects the output voltage.

**Materials:**

> 1 100 Ω resistor
>
> 1 breadboard
>
> 1 9 V battery
>
> 1 DMM
>
> Jumper wire
>
> 1 photocell

**Instructions:**

1.  Starting from the setup from the previous lab, disconnect the battery.

2.  Replace R2 with the photocell.

3.  Reconnect the battery.

4.  Move your hand to cover the photocell and watch how the voltage out ($V_{out}$) changes on your DMM display. This is because the value of the resistance of the photocell changes depending on the amount of light it is exposed to.

R1 = 100 Ω

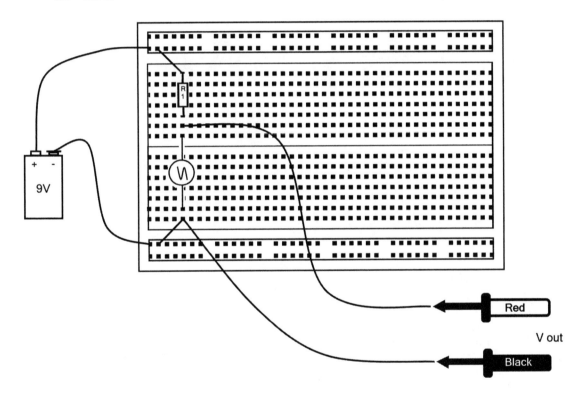

# Capacitors

## In This Chapter

- Using capacitors to store charge
- Understanding the inner workings of capacitors
- Getting acquainted with electrolytic, nonpolarized, and variable capacitors
- Charging and discharging capacitors

*Capacitance* is the ability to store charge (electrons). A *capacitor*, also sometimes called a condenser, is an electrical component that can store charge. Capacitors are used for many purposes, including smoothing out spikes or drops in voltage and releasing quick bursts of power in applications such as a camera flash.

## How a Capacitor Works

A capacitor is constructed of two layers, or plates, of conductive material separated by an insulator. When voltage passes through a capacitor, it creates an electric field in insulating material, called the dielectric, situated between the two layers. The field holds an electric charge in the dielectric.

When voltage is applied, the current travels through the capacitor and the dielectric moves electrons from one of the conductors to the other creating opposing polarity—one layer with a deficiency of electrons so it has a positive charge and the other layer with an excess of electrons yielding a negative charge. The net charge of the capacitor is zero.

When current flows through the capacitor, it charges the capacitor so that the voltage matches the source voltage. Once the capacitor is charged, the voltage that flows through the capacitor is the same as the voltage that enters it.

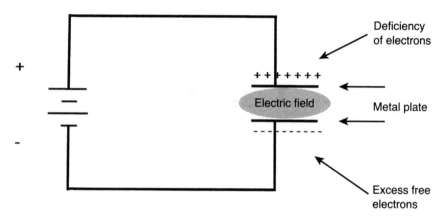

*The parts of a capacitor. Note that the dielectric is the area between the two layers that holds the electric fields.*

A capacitor maintains a constant voltage, and the process is often called *smoothing the current*. It accomplishes this by using its storage capacity—its capacitance—to either donate or accept electrons when there is a drop or a spike in voltage.

**HIGH VOLTAGE!**

Capacitors can be dangerous to work with. In some common electronic devices, capacitors hold enough charge to be fatal, and they can hold a charge for quite some time after the device it is connected to no longer works. Always assume that capacitors are fully charged and can release their full charge.

If there is a voltage drop in a circuit, the capacitor will push electrons toward the direction of the source voltage, acting as a voltage source. This may seem counterintuitive because we are used to thinking of electrons flowing in a single direction, from the positive to the negative terminals of a power supply. But electrons also flow "backwards" in a circuit if the charge in the capacitor is positive compared to the voltage entering the capacitor. Electrons are still flowing positive to negative, but until the charges match, the capacitor acts as source voltage until it has exhausted its capacitance.

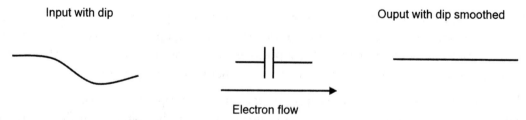

*A capacitor can act as a source when there is a dip in power.*

If there is an increase in voltage from the voltage source, the capacitor will resist that change by storing the excess electrons up to its rated capacitance. When this happens, a capacitor is said to be acting as a load, and the incoming voltage is reduced.

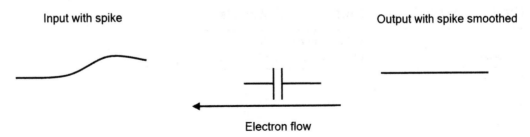

A capacitor can act as a load when there is a spike in power.

# Farads

Differences in the dielectric medium determine the amount of charge that can be held in the electric field of the capacitor. The measure of capacitance is the farad, represented by an F. A farad is equal to the current in *coulombs* required to raise the voltage across the capacitor by one volt.

**DEFINITION**

A **coulomb** is an ampere second, the amount of current provided by one ampere of power in one second.

The most commonly used units of capacitance are the microfarad ($\mu$F, $1 \times 10^{-6}$), nanofarad (nF, $1 \times 10^{-3}$), and the picofarad (pF, $1 \times 10^{-4}$). Integrated circuits use even smaller capacitors, which require an even smaller unit of measurement, a femtofarad (fF), which is $1 \times 10^{-15}$ F.

# Relative Permittivity

Different dielectric materials provide varying levels of capacitance, as shown in the table on the next page. Remember that the dielectric is the insulating material in a capacity that stores charge. Permittivity is the ability to create an electric field. Permittivity is measured relative to a pure vacuum as the dielectric; this is called relative permittivity.

A higher relative permittivity means that a capacitor can store more charge; a lower relative permittivity would hold less charge. A pure vacuum has a relative permittivity of 1.0. The relative permittivity of a material is also given as its dielectric constant.

### The Dielectric Constant for Some Materials

| Material | Relative permittivity (dielectric constant) at 0°C |
|---|---|
| Vacuum | 1.0000 |
| Air | 1.0006 |
| Teflon (PTFE)* | 2.0 |
| Polypropylene | 2.20 to 2.28 |
| Polystyrene | 2.45 to 4.0 |
| Transformer oil | 2.5 to 4 |
| Hard rubber | 2.5 to 4.80 |
| Silicones | 3.4 to 4.3 |
| Bakelite | 3.5 to 6.0 |
| Quartz, fused | 3.8 |
| Glass | 4.9 to 7.5 |
| Porcelain, steatite | 6.5 |
| Distilled water | 80.0 |
| Hydrofluoric acid | 83.6 |
| Titanium dioxide | 173 |
| Strontium titanate | 310 |
| Barium strontium titanate | 500 |

*polytetrafluoroethylene

## Capacitor Ratings

Beyond the materials used to create a capacitor, other factors need to be considered when choosing a capacitor. Each designated value for a capacitor also has tolerance, with the performance varying some percentage higher or lower. In addition, higher or lower temperatures can change the permittivity of the dielectric. Capacitors have maximum circuit voltages, so even if the capacitor has the proper capacitance it won't be appropriate for your particular circuit if the voltage of the circuit exceeds that maximum voltage. Each of these factors needs to be considered when choosing a capacitor.

## Nominal Value and Tolerance

Capacitors have a code imprinted on them, which represents their nominal value. The code used to indicate the capacitor's nominal value is similar to the Resistor Color Code (see Chapter 7). The first two digits are the significant digits (the digits to the left of the decimal point), whereas the third digit is the exponent value (the powers of ten). As most capacitors have very small values, the code expresses values in pF.

**WATTAGE TO THE WISE**

Electronics buffs frequently refer to capacitors as *caps* and picofarads as *puffs*. Don't be surprised to hear seasoned electronics folk toss in several other odd terms. When you're not familiar with the jargon, be sure to ask questions.

Immediately following the two- or three-digit nominal value code, you will often find a single letter representing the tolerance of the capacitor. The following table matches each code with its respective tolerance:

### Tolerance Codes for Capacitors

| Letter symbol | Tolerance—% |
|---|---|
| B | ±0.10 |
| C | ±0.25 |
| D | ±0.5 |
| E | ±0.5 |
| F | ±1 |
| G | ±2 |
| H | ±3 |
| J | ±5 |
| K | ±10 |
| M | ±20 |
| N | ±0.05 |
| P | +100−0 |
| Z | +80−20 |

Not all capacitors are marked with a tolerance code. A tolerance of ±10 percent means that the value is somewhere in the range of 10 percent higher or lower than the stated value.

## Temperature Coefficients

A capacitor's performance can vary with the operating temperature. Capacitors are sometimes marked with a temperature coefficient, usually given as parts per million (ppm)/°C. The temperature coefficient of 100ppm/°C is equivalent to .01 percent/°C.

Temperature coefficients are also usually assigned a negative or positive value. A positive temperature coefficient means that there is a positive correlation between temperature and capacitance; in other words, an increase in temperature increases the capacitance and a decrease in temperature decreases the capacitance. A negative temperature coefficient means that the capacitance decreases when temperature increases and vice versa. There is a negative correlation between temperature and capacitance.

## Breakdown Voltage or DC Working Voltage

At certain voltages, capacitors can break down. The electric fields formed can make the dielectric fields conductive, resulting in the capacitor being unable to hold a charge. The peak operating voltages for proper long-term operation of capacitors are given as the direct current working voltage (DCWV). Another value that may be given is the surge voltage, which is the peak voltage that a capacitor can withstand for temporary surges or drops. Usually, the higher the resistance value, the lower the working voltage is.

# Polarized Capacitor Types

Some capacitors are designed to be operated in just one direction in a circuit. They usually can only be used in a circuit with direct current (DC) and will break down if improperly inserted.

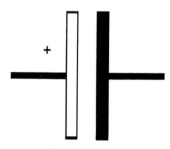

Polarized capacitor
symbol

*Circuit diagram symbol for a polarized capacitor.*

## Electrolytic Capacitors

Electrolytic capacitors are a type of polarized capacitor that uses an ionized conductive liquid as one of the conductive layers or plates. An aluminum electrolytic capacitor is made of two layers of aluminum foil, with one of the layers covered in a thin film of oxide separated by a paper insulator soaked in an electrolyte solution. The coated layer and the soaked paper insulator form a *cathode* (from which electrons flow), whereas the noncoated layer acts as the *anode* (into which electrons flow).

**DEFINITION**

An **anode** is an electrode that current flows into. A **cathode** is an electrode that current flows out of. Some common mnemonics to help remember these terms are ACID (Anode Current Into Device) for anode and CCD (Cathode Current Departs) for cathode.

Electrolytic capacitors provide higher capacitance than nonpolarized dielectric capacitors (especially in relation to their volume), but they have a lower breakdown voltage and life span. Electrolytic capacitors are primarily used in higher voltage situations such as power smoothing and AC applications.

## Tantalum Capacitors

A tantalum capacitor is a specific type of electrolytic capacitor. The tantalum capacitor was developed by researchers seeking a more stable capacitor that is still high capacitance relative to its volume. Tantalum (a lustrous metal element) powder is formed into a pellet and coated in an oxide layer. Then a conductive material or an electrolytic solution "plate" surrounds the pellet and the oxide layer. The result is a polarized capacitor that has high capacitance with higher breakdown voltages and a potential life span of decades.

Tantalum resistors are used in many compact devices or in situations where long-term reliability under high-temperature conductions is necessary.

# Nonpolarized Capacitor Types

There are many types of nonpolarized capacitors created for various uses and operating environments. The dielectric materials used depend on factors including voltage environments, cost, size, tolerance, and life span. The table on the next page lists some common dielectric materials and their properties.

| Dielectric | Properties & Notes |
|---|---|
| Paper | Used in high-voltage situations; being replaced by plastic-film capacitors |
| PET* film | Used in high-voltage and high-temperature situations; has replaced most paper types |
| Polystyrene | Good general use, slight negative temperature coefficient; only good for midtemperature applications |
| Polypropylene | One of the most popular general-purpose caps; more subject to damage from voltage |
| Teflon (PTFE**) | Very reliable and high-temperature performance; downsides are high cost and large size |
| Glass | Very stable, very reliable; high cost |

*\* Polyethylene terephthalate*

*\*\* Polytetrafluoroethylene*

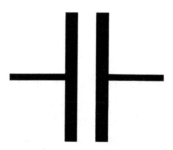

Nonpolarized capacitor
symbol

*Circuit diagram symbol for a nonpolarized capacitor.*

# Variable Capacitors

Some capacitors can be adjusted to increase either the area between the plates or the amount of overlap between the plates. Applications of variable capacitors include digital tuners, sensors in industrial applications, and capacitor microphones, which adjust their volume according to the effect of sound on a diaphragm.

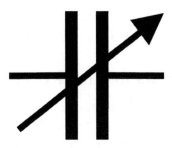

Variable capacitor
symbol

*Circuit diagram symbol for a variable capacitor.*

# Capacitors in a Circuit

Unlike with resistors, when you combine capacitors in a series circuit, the sum of the capacitance is less than the capacitance of the individual capacitors. The formula used to calculate capacitance in a series is shown here (note that this is the same formula used for resistors in parallel):

$$\frac{1}{C_T} = \frac{1}{C_1} + \frac{1}{C_2} + \frac{1}{C_3}$$

*Calculating capacitance in series, where $C_T$ represents total capacitance and $C_1$, $C_2$, and so on, represent the first and the second capacitors in the series.*

When you combine capacitors in parallel, the total capacitance is the sum of the individual capacitances. Again, this is the opposite of the behavior of resistors in parallel.

$$C_{total} = C_1 + C_2 + C_3 ... + C_n$$

*Calculating capacitance in parallel.*

When a circuit first has current flowing through it, a capacitor has not yet charged. A capacitor will let current flow through as it charges. A fully discharged capacitor acts as a short circuit; there is no voltage drop as the current flows.

When a capacitor is fully charged, the voltage of the capacitor matches the source voltage and current will not flow. In other words, a fully charged capacitor acts as an open circuit.

## Transient Time of Capacitors in a DC RC Circuit

For a capacitor to function in a circuit it requires some resistance provided either by a resistor or a load in addition to the power supply. To describe the action of a capacitor, we consider it as part of a simple DC circuit called an RC (resistor-capacitor) circuit.

The time between when a capacitor is fully discharged to when it's fully charged (or fully charged to fully discharged) is called the transient time.

The charging process isn't steady; when the switch is first closed to start the flow of current, there is an initial burst of charging and then a slower increase until the capacitor is said to be fully charged. The time it takes to charge/discharge is calculated by determining the time constant.

The time constant (symbolized by •) is determined by the resistance and the capacitance in a circuit. The formula is • = R × C where R is resistance given in ohms (Ω) and C is capacitance given in farads (F).

In the first time constant of charging, the capacitor is charged to approximately 63 percent. During each time constant, it moves closer to being fully charged. Please see the following chart.

| Time | Charge |
|------|--------|
| 1RC | ≈63% |
| 2RC | ≈86% |
| 3RC | ≈95% |
| 4RC | ≈98% |
| 5RC | ≈99% |

*The transient time from zero to approximately fully charged.*

You can see that it takes five time constants to reach close to a full charge. This is an approximate value because the capacitor never approaches being fully charged or fully discharged. It continues to increase ever so slightly toward a charge of 100 percent or 0 percent when discharging, but for almost all purposes, five time constants will bring the capacitor to a steady state (either fully charged or fully discharged).

The increase in voltage across the capacitor as it goes from 0 V to the source voltage value, and then discharging back to zero can be graphed as a curve known as an exponential curve.

## The Least You Need to Know

- Capacitors store charge. Capacitance is measure in farads.
- When working with capacitors, you should always assume that the capacitor is fully charged and could potentially cause injury or damage the circuit and its components.
- Capacitors are of two conductive plates with an insulating layer, called the dielectric, between them. The type of material used for the dielectric affects the capacitance.
- Capacitors are either in steady state (fully charged or fully discharged) or are in transient state. You can determine the charging (or discharging) time if you know the current and the resistance in the circuit.

## Lab 8.1: Charging Capacitance

In this lab you use a resistor and an LED to show the charging of a capacitor.

**Materials:**

>  9 V battery
>
>  Breadboard
>
>  1 470 Ω resistor (yellow violet brown)
>
>  1 220 µF electrolytic capacitor (or cap)
>
>  1 red light-emitting diode (LED)
>
>  Jumper wire

**Instructions:**

1.  Place the components on the breadboard as indicated on the diagram. Note that the resistor doesn't have polarity so it can be connected to the breadboard in either direction. The LED needs to be connected to the resistor on its positive lead. (Note: The flat side of the LED is the – side.)

2.  Connect the capacitor at its positive lead. (Note: the cap has + and – marked, and the longer lead is +.)

3.  Connect the jumper wires to the battery. The LED lights up for about one second, then slowly dims. This is the capacitor charging.

To repeat the lab, first disconnect the battery, then discharge the capacitor by touching a jumper wire between the two leads of the capacitor. Then start again.

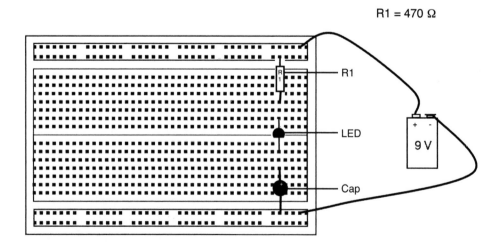

# Diodes

## In This Chapter

- Using diodes to allow current to flow in a single direction
- Classifying diodes by their voltage drop, peak inverse voltage, and recovery time
- Distinguishing between P-N diodes, Zener diodes, Schottky diodes, light-emitting diodes, and photodiodes
- Using diodes as power rectifiers

Diodes are components that permit electricity to flow in a single direction and act against its flow in the opposite direction. They have two terminals: a positive terminal, also called an anode, and a negative terminal, or cathode. While this seems like a pretty basic function, diodes have been used to do some revolutionary things.

Diodes are the technology that made early radio transmission and reception possible. They are essential in converting alternating current (AC) to direct current (DC). They prevent damage to electronics from high voltages. The use of diodes to create logic gates makes possible the computations that are the backbone of computers. Your digital camera's flash depends on diodes. All of these functions are based on the ability of a diode to determine when to let current through, when to block current, and when it will send a big jolt of current back through a circuit.

Diodes are primarily used in small-signal applications (1 A or less) and to convert AC to DC power, a process called power rectification. Diode arrangements for power rectification are usually referred to as *rectifiers*.

Diodes can also be used to protect against spikes or other power fluctuations that may cause a reverse current. When used in this capacity, diodes are called *transient protectors*.

Another common use of diodes is as reverse polarity detectors in electronic devices to protect the device against connecting to a power supply in which the polarity reversed.

# How Diodes Work

Most diodes are constructed by stacking a p-type semiconductor, which is positively charged; a boundary (or depletion) region called a *P-N junction*, which is neutrally charged; and an n-type semiconductor, which is negatively charged. This creates a valve-like component that conducts current in a direction from the p-type side (called the *anode*) to the n-type side (called the *cathode*), but not in the opposite direction.

**WATTAGE TO THE WISE**

Although the comparison is far from perfect, it might be helpful to think of diodes as the electronic equivalent of check valves, which are plumbing fixtures that allow water to flow in just one direction.

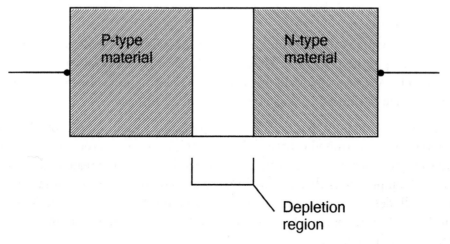

*A semiconductor diode.*

**WATTAGE TO THE WISE**

Most modern diodes are made using semiconductors, but for some high-voltage situations a vacuum tube is used instead.

Before the diode can function properly it first needs to be "turned on." When current is first applied, there is a forward voltage drop that needs to be overcome. This is the amount of current that is necessary before current can pass through the diode. This forward voltage drop is .7 V in diodes made of silicon and .3 V in germanium diodes.

Until this forward voltage drop is overcome, the diode acts as an insulator (it doesn't permit any current to pass through) even if it is of the correct polarity. When the diode does overcome the forward voltage drop, it is then "turned on" and only allows the passing through of current in one direction. This direction is called the *bias* of the diode. The time it takes to overcome the forward voltage drop is called the recovery time.

**DEFINITION**

**Bias** refers to the direction of the voltage, either forward or reverse.

The direction the current travels is determined by its polarity (its positive or negative charge). Electrons flow from the positive to the negative. The two terminals of the diode are labeled as "a" for anode (positive side) and "k" for cathode (negative side). The diode acts like a one-way valve to the flow of electric current. Current flowing from anode to cathode flows with ease but current flowing from cathode to anode is blocked.

When the diode encounters current of the opposite polarity, the diode acts as an insulator. It is this ability to block current in one direction and allow it in the opposite direction that gives diodes so many uses.

A diode's insulating effect is not limitless. It is subject to a peak inverse voltage (PIV). This is the highest amount of voltage in the reverse bias the diode can withstand before failing. This is like the amount of pressure that the valve can stand before bursting. Once PIV is exceeded, the number of electrons can overwhelm the P-N junction. This is called *breakdown*. Usually breakdown occurs in higher values, typically 50 V or more.

**HIGH VOLTAGE!**

Be sure to pay close attention to your diode's PIV rating to avoid breakdown. In most cases, if breakdown occurs, the component containing the diode and any other components on the circuit will be destroyed.

# Types of Semiconductor Diodes

Which type of semiconductor is used in a diode and how the diode is constructed determine its specific function in an electronic device. The following sections describe some common semiconductor diodes you might encounter in various electronic devices.

## Common Silicon Diodes

The most common types of diodes are silicon diodes, also called P-N diodes. They are usually enclosed in a glass tube, with a stripe indicating the cathode (negative) terminal. They have two leads, or wires, that connect into the circuit.

*A silicon diode. The stripe indicates the location of the cathode within the glass tube.*

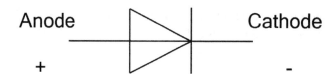

*The symbol for a diode on a circuit board diagram. It refers to any common diode; if a particular diode construction is required it will be represented by that specific symbol.*

## Zener Diodes

A Zener diode depends on a precise PIV to act as a type of voltage-dependent switch. These are sometimes called constant voltage diodes. When the PIV is reached, current can travel in the reverse direction. It is a type of avalanche diode, which is any kind of diode that depends on a surge in voltage for reverse current flow.

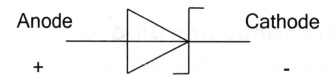

*The symbol for a Zener diode on a circuit board diagram.*

Zener diodes are used in surge protectors, to protect against spikes. They are usually used with a resistor to limit the current so the maximum current parameters are not exceeded.

## Schottky Diodes

Schottky diodes are constructed with a metal-semiconductor design. A conducting metal and a metal-oxide film are attached to an n-type semiconductor. This construction provides both a very low voltage drop (typically between .15 V and .45 V) and a very fast recovery time.

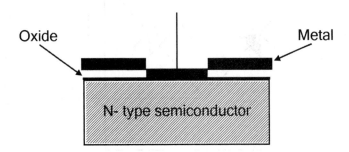

*How Schottky diodes are constructed.*

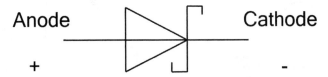

*The symbol for a Schottky diode on a circuit board diagram.*

Schottky diodes are often used as power rectifiers, but their properties also make them ideal components in radio-frequency circuits and other specialty applications.

# Power Rectifiers

As noted at the beginning of this chapter, diodes can be arranged to act as AC-to-DC power rectifiers. A bridge rectifier is a type of rectifier consisting of a series-parallel arrangement of diodes that ensure that only one polarity of current is the output no matter the polarity of the input current.

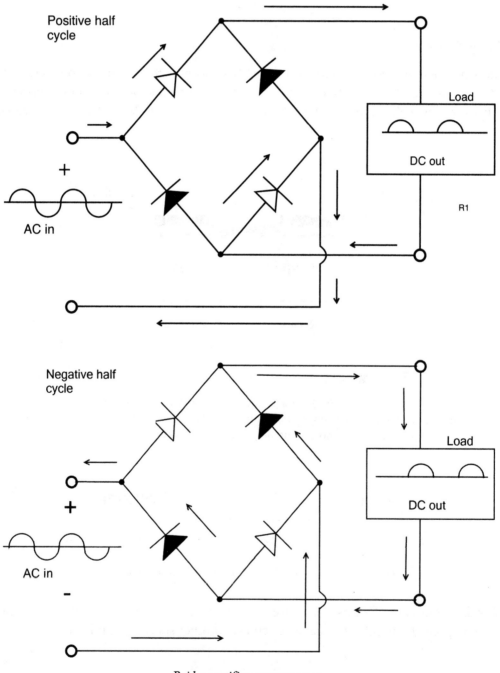

*Bridge rectifier arrangement.*

This may seem like a very complicated construction, but if you want to know how an AC-to-DC rectifier works it is important to work through the two images. In the first figure, during the positive half cycle of the AC wave, the current flows through the white diodes (forward biased) and DC flows out of the rectifier.

In the negative half cycle of the AC wave (second figure), the current travels through the black diodes (reverse biased) and is returned through the rectifier to the AC source instead of passing to the DC out. This means that only current in the correct direction (DC) passes through the bridge rectifier.

## Light-Emitting Diodes (LEDs)

An LED is a P-N junction diode constructed of materials that are considered direct band-gap materials. In these specialized materials, when the electrons combine with the holes, energy in the form of a photon is released. The type of semiconductor material used in the LED determines the frequency or wavelength of the light emitted and thus the color.

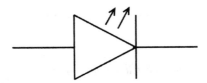

*The symbol for an LED diode on a circuit board diagram.*

To produce white light, you can use a combination of other colored LEDs or a phosphor coating. LEDs in the infrared spectrum are used in many remote-control devices. LED lighting provides a very efficient source of light output per power input, especially in low-power situations.

LEDs can be connected in series to power several LEDs off of one power source. Connecting LEDs in parallel is usually not necessary because the same current required to power each LED in parallel could be provided in series, whereas in a parallel arrangement each LED would need its own resistor.

## Photodiodes

A photodiode is a specialty diode that detects light. In other types of diodes, the P-N junction area is usually shielded from light, so light energy doesn't interfere with the function of the diode. A photodiode is designed to register light energy and so is not shielded from light.

Photodiodes use a PIN structure (p-type semiconductor, *i-type semiconductor*, n-type semiconductor) or a NIP structure (n-type semiconductor, i-type semiconductor, p-type semiconductor). This forms a wider band so the diode can detect longer light wavelengths.

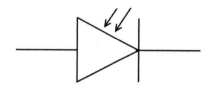

*The symbol for a photodiode on a circuit board diagram.*

Photodiodes have many electronics applications, from optical storage and retrieval to telecommunications and photography.

# Handling Diodes

Handle diodes with care. It is usually a good idea to use your multimeter to test your diode to ensure that it is operating before soldering it into your circuit.

Always pay attention to the polarity of the leads, as diodes function correctly only when properly polarized. The positive lead (anode) may be labeled with an "a" and is slightly longer. The negative lead (cathode) may be labeled with a "k" and is slightly shorter. As these differences are small and usually require a magnifying lens to read; a better method of telling the anode from the cathode is to look for the flat side of a round diode assembly. The flattened side is the cathode. We recommend looking for at least two of these three features before soldering.

As diodes are easily damaged by heat, extra care should be taken when soldering them. It is usually a good idea to use a heat sink clipped to the lead between the soldering joint and the diode body.

## The Least You Need to Know

- Diodes are electronic components that allow current to flow in a single direction and resist the flow of current in the reverse direction.

- Some key concepts to understand are voltage drop (the loss of voltage that is required to "turn on" the diode); peak inverse voltage (PIV, the highest amount of reverse voltage a diode can block); and breakdown (the avalanche of reverse voltage that overtakes a diode after PIV is exceeded).

- Most diodes are made using semiconductors and include P-N diodes, Zener diodes, Schottky diodes, power rectifiers, light-emitting diodes, and photodiodes. The type of semiconductor used determines its function.

# Transistors

## In This Chapter

- Using transistors to amplify signals
- Distinguishing between bipolar junction transistors and field effect transistors
- Applying voltage to transistors and the effects
- Looking at specialized FETs developed for cutting-edge applications

A transistor is an electronics component that amplifies signals or acts as a switch in a circuit. Transistors enable us to control the amount of current that flows through a circuit, diverting it or even reversing it.

Prior to the invention of the transistor in the first half of the twentieth century, electronics were dependent on vacuum tubes, which were bulky, expensive to produce, not very rugged, required a lot of power, and gave off quite a bit of heat.

The invention of transistors led to the field of modern electronics and made possible smaller and more powerful electronic devices because transistors require little power, don't generate a lot of heat, and are very reliable. Tiny transistors can be used in complicated integrated circuits, making it possible to pack a lot of computing power in a small package.

There are two main types of transistors: bipolar junction transistors (BJTs) and field effect transistors (FETs).

## Bipolar Junction Transistors (BJTs)

BJTs are basically two diodes put end to end. As you learned in Chapter 9, diodes allow the flow of current in only one direction. If the polarity is properly matched, there is a small drop in voltage and the current passes through. If the polarity is not matched, the diode stops the current (unless, of course, it exceeds the peak inverse voltage and breakdown occurs). But if a transistor is two diodes put end to end, you would think that each diode would block current in the opposite direction, allowing neither to pass. That isn't the case.

So why doesn't a transistor act as an insulator in both directions? As with diodes, resistors have leads connecting to the collector (C) (which is analogous to the anode on a diode) and the emitter (E) (which is analogous to the cathode on a diode). Unlike diodes, transistors have a third lead, called the base lead, which connects to the base (B). This base lead provides a voltage, which changes everything. More on that in a bit; first, let's take a closer look at the two types of BJTs—PNP and NPN—and how they are constructed.

The difference between PNP and NPN BJTs is the bias, which is the direction of current which is allowed to pass.

In a PNP BJT, there are two positively charged regions—the collector and the emitter—each with excess holes. The depletion region in between the two is negatively charged in that it has excess electrons. The borders between the regions are called junctions. The base-emitter junction is reverse-biased, and the base collection junction is forward-biased. A small voltage is applied to the base region.

In an NPN BJT, there are two negatively charged regions—the collector and the emitter—each with excess electrons. The depletion region is positively charged in that it has excess holes. In an NPN BJT, the base-emitter junction is forward-biased, and the base collection junction is reverse-biased. A small voltage is applied to the base region.

## How Amplifiers Work

What does the addition of a voltage at the base of NPN and PNP BJTs accomplish? Let's look at the operation of an NPN BJT as an example.

Voltages are applied at the collector end and to the base. If the current is of the same polarity—negative in our case—it will pass through. Electrons flow by combining with the holes, but as they move, new holes open up behind them. While an n-type semiconductor has excess electrons, holes are still being created, pulling in more elections to combine. The overall polarity remains negative (because in an NPN the two negatively charged regions are the majority carrier), but holes are still being created (in the positive region known as the minority carrier in an NPN). When current moves in the same direction as the majority carrier, current passes through and the depletion zone is made very thin, allowing even faster transmission of electrons.

The base voltage acts like a pump in that it moves some electrons from the emitter region into the base circuit, leaving behind holes in the base region. The pumping action increases the voltage across the base-emitter junction. This creates a greater push through the majority carrier from collector to base and on to the emitter, which increases the overall voltage (which acts as a push) and amplifies the current that can travel through the transistor.

In BJTs, a small voltage applied to the base lead creates a much larger current (an exponential increase) flowing from the collector anode (+) to the emitter cathode (-). The emitter cathode is usually shown by an arrow on the component diagram. Notice that the direction of the arrow indicates whether the transistor is PNP or NPN and indicates the direction of current flow.

PNPs are reverse-biased. NPNs are forward-biased.

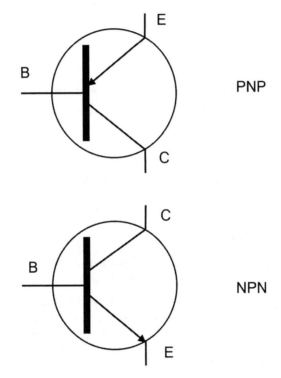

*Circuit diagram symbols for a PNP BJT and an NPN BJT.*

## BJTs Under Varying Voltages

A transistor's functions depends on the voltages applied and the polarity of the current. The following formulas show the function of the transistor in various scenarios. The subscripts refer to the base ($V_B$), the collector ($V_C$), and the emitter ($V_E$).

If $V_E < V_B < V_C$, then the transistor acts as an amplifier.

If $V_E < V_B > V_C$, then the transistor is a conductor.

If $V_C > V_B < V_E$, then the transistor is an open switch and cuts off the flow.

## Gain

The amplification effect in a transistor is called a *gain*. The equation representing the gain is:

Voltage gain = $V_{OUT}$ / $V_{IN}$.

The gain is represented by $\beta_{FE}$ or by $h_{FE}$. There is a proportional relationship between the gain and the base current, represented as:

$\beta_{FE} = I_C / I_B$.

That is, the gain in current over a transistor is the current at the collector divided by the current at the base.

# Darlington Pairs

Darlington pairs are two transistors sold as one with the leads arranged as if they are a single unit. This arrangement allows for much more significant amplification and is used in situations where gains in the order of 1,000 are called for.

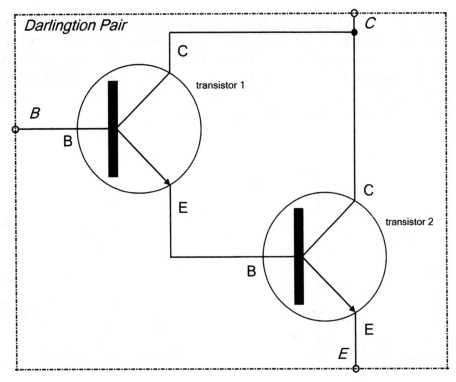

*Darlington pairs.*

# Field Effect Transistors

FETs have four terminals or leads: gate, drain, source, and body. The body is the substrate of the transistor; the gate is analogous to the base on a BJT-type transistor; the drain functions the same as a BJT collector; and the source does the same work as a BJT emitter.

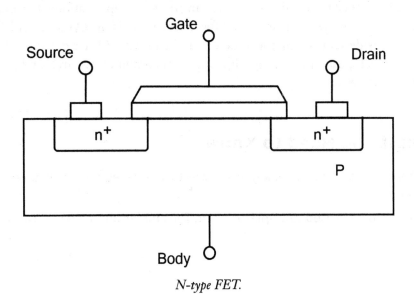

*N-type FET.*

The gate opens and shuts depending on the voltage applied. The source to drain area is the channel through which current travels when the gate allows connection between the two areas. There are two general types of FETs: enhancement or depletion mode. These modes are further categorized by their bias (n-type or p-type). Further classification is made based on the materials used or by the transistor's function. MOSFET is a metal-oxide-semiconductor FET that is used in digital logic gates.

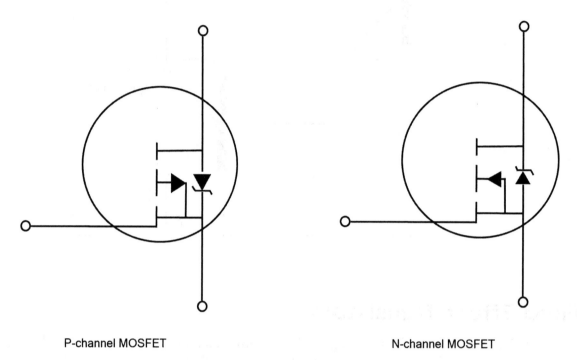

P-channel MOSFET                    N-channel MOSFET

*The circuit diagram symbols for MOSFETs.*

There are many other FETs adapted for use in particular applications. For example, an ISFET is an ion-sensitive FET; it is used for detecting ions in solution. An EOSFET uses an electrolyte-oxide semiconductor to detect brain activity. Other cutting-edge FET types include CNTFETs, used in cutting-edge quantum computer research and DNAFETs, used in DNA sequencing.

## The Least You Need to Know

- Transistors are one of the key inventions that powered the development of modern electronics.
- Transistors are used to amplify current. The amplification effect is called gain.

- Bipolar junction transistors (BJTs) are basically two diodes placed end to end; it is the addition of a voltage at the base that makes a transistor into an amplifier. The key parts are the collector, the emitter, and the base.

- Field effect transistors (FETs) have four key parts: the body, the gate, the source, and the drain.

- There are many specialized FETs used in medical, chemical, and quantum computing applications.

## Lab 10.1: Using a Transistor to Amplify Current

To see a transistor in action, let's see if we can overcome the very significant resistance offered by the human body when it is put between a 9 V power source and an LED requiring 20 mA (.020 A) of current to operate.

**Materials:**

> 1 9 V battery
>
> Breadboard
>
> 1 2N2222 NPN transistor
>
> 1 red LED
>
> 1 560 Ω resistor (green blue brown)
>
> Jumper wire

**Instructions:**

1. Connect the components as shown in the diagram.

2. Use Ohm's Law to calculate the current provided by the circuit without the use of a transistor. The resistance of the human body when dry is around 100,000 Ω. The LED has a voltage drop of 2 V. So I = (9− 2)/100,000 + 390 (I = V / R). I = .00007 A (.7mA).

3. The LED requires 20 mA of current to light, so without an amplification of current we wouldn't be able to light the LED.

4. Touch the two leads, each in one hand.

5. Even with the addition of your body's resistance to the circuit, the lamp will light due to the transistor amplifying the current to the required 20 mA.

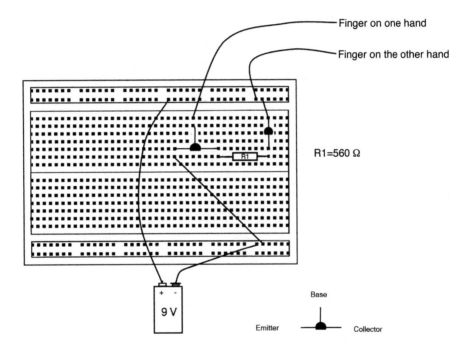

Finger on one hand

Finger on the other hand

R1=560 Ω

9 V

Base

Emitter    Collector

# Power Sources and Power Supplies

## In This Chapter

- Making sense of the many different types of power sources and supplies
- Creating a chemical reaction in a voltaic cell to produce energy
- Turning electricity from a main power source into usable power
- Getting acquainted with specialized power supplies

This chapter is all about power—specifically, *power sources* and *power supplies*. Although the two terms sound quite similar, they have distinct meanings.

A power source provides electricity. In electronics, we work with two primary power sources: direct current (DC) voltage from stand-alone batteries and alternating current (AC) power, also called *AC mains*, provided by the electric utilities. Each of these sources can provide electricity to a power supply.

A power supply is the power source in a circuit for a particular electronic device, whether that source is a battery, a power outlet, or a small circuit that adapts power from an outlet to the appropriate voltage needed for the device.

# Batteries

Batteries—from the tablet-size cells that power our watches and cell phones to the large units in our cars—are stand-alone power sources that don't rely on electricity from wall outlets. Instead, they produce electrical energy through a chemical reaction.

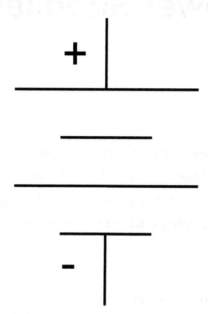

*The symbol for a battery in a circuit diagram.*

## How Batteries Produce Energy

A battery consists of one or more voltaic cells. Voltaic cells usually consist of two half cells, each with an electrode surrounded by an electrolyte and separated by a separator. The separator allows for the movement of ions, but not the mixing of materials, between the electrodes or the electrolyte. In the course of the chemical interaction (called a reduction and oxidation chemical reaction) between the electrolytes in the two half cells, the charged ions are constantly moving until the reactions between the two electrodes are completed, equilibrium is reached, and a battery is created.

To function as a power source, an external connection to a circuit is made between the two electrodes. Negatively charged ions travel through the electrolyte to the negative electrode and positively charged ions travel through the electrolyte to the positive electrode. The movement of ions creates an electrical potential difference in energy between the two electrodes; this potential difference is voltage. One of the electrodes is a positive terminal and the other is a negative terminal. When connected to a circuit, the voltage pushes current from the negative terminal through the circuit and returns to the positive terminal.

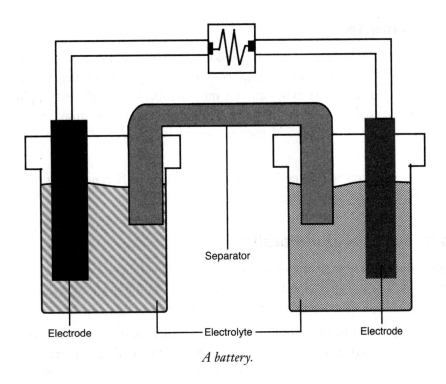

*A battery.*

## Primary vs. Secondary Batteries

Batteries that are designed to produce energy until the chemical reaction is depleted are called *primary batteries*. Once primary batteries lose their charge, they cannot be recharged to produce more energy.

Batteries that can be recharged are called *secondary batteries*. Secondary batteries can be recharged by introducing a reverse voltage into the cells. For example, a car battery can be recharged through the alternator as it runs, and rechargeable AA batteries can be recharged through their recharging station plugged in to a household outlet. The ability to recharge a secondary battery isn't limitless; eventually the ability for the battery to achieve full charge diminishes.

Many portable electronic devices use rechargeable dry cell batteries. These are made with a paste instead of a liquid electrolyte. One common dry cell rechargeable battery construction is the lithium ion type.

**TITANS OF ELECTRONICS**

The Italian physicist Alessandro Volta (1745–1827) is credited with creating the first working electrochemical voltaic cell. He was made a count by Napoleon in 1910, and the Voltian Temple was created in his honor in a museum in Como, Italy.

## Voltages in Batteries

Each cell in most common battery types produces 1.3 to 3 V; but across most battery types, the voltage is usually 1.5 V. The chemical construction of each cell determines its voltage. This is why some battery types are not interchangeable in different applications. For instance, because rechargeable AA batteries made with nickel cadmium produce 1.25 V and alkaline AA batteries produce 1.5 V, they cannot be used together to power an electronic device.

Multiple cells are often contained in a battery to produce different voltages. When you see a 9 V battery, think of it as having six cells (6 × 1.5 V = 9 V).

## Batteries in Series and in Parallel

To obtain higher voltages than a single battery can provide, you can connect multiple batteries in series. Three batteries stacked in a standard flashlight are in series; together they provide more voltage than each individual battery can.

A parallel arrangement of batteries does not provide increased voltage. However, when batteries are connected in parallel, they reduce the resistance of the batteries; this reduced resistance results in an increase in the overall current. Generally, the larger the battery is, the larger the resistance. You can see parallel battery arrangements in many older electronics, such as portable radios.

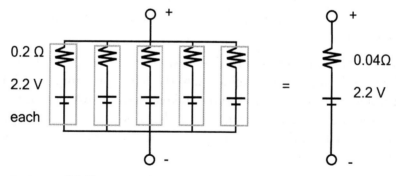

*Batteries in parallel. You can see that the resistance of each battery, .2 Ω each, is reduced by this arrangement to an overall resistance of just .04 Ω. The voltage overall is not increased, it remains at 2.2 V.*

## Amp-Hours

Batteries are usually categorized by their voltage, but they are also categorized by their amp-hours (AH). An amp-hour is a unit of energy capacity; it is equal to the amount of continuous current in amps per hour that a battery can provide before depletion. You will also see energy capacity described in milliamp-hours (mAH).

The discharge time for a battery depends on the load on the circuit. Lighter loads take longer to discharge than higher loads. The ratings given by battery manufacturers vary according to operating conditions (temperature, age of the battery, amount of heat lost to resistance in a circuit, and so on), but you can get approximate hours of operation by dividing the amp-hour rating by the continuous current (in amps).

**WATTAGE TO THE WISE**

Several types of nonchemical batteries, including solar cells and fuel cells, have been developed. Solar cells collect electrons that are dislodged from a semiconductor by the action of the photoelectric effect. Fuel cells collect the energy released by the oxidation of chemicals outside of the cell, such as hydrogen and oxygen. Both of these technologies produce long-lasting voltaic cells without the harmful waste products associated with metal/acid batteries.

# AC-to-DC Power Supplies

Very few items in a typical household run directly on the AC power provided from the outlet. Usually only motorized high-power appliances such as refrigerators and washing machines are designed to run on AC. Most everything else converts the current into DC.

Power supplies connect via a power cord to the AC mains. The power supply then goes through the following steps to produce the DC at the right voltage to power the appliance:

1. The transformer steps down (reduces) the high-voltage AC to lower-voltage AC.

2. A rectifier converts AC power to DC power.

3. The DC wave is smoothed and regulated to as close to a straight-line wave as possible.

If you look around at the various power cords in your home or office you might see what is commonly called a *wall wart* or an AC adapter; each of these have a transformer, rectifier, some smoothing of current, and a regulator. Your cell phone, computer, and stereo system each have their own power supply.

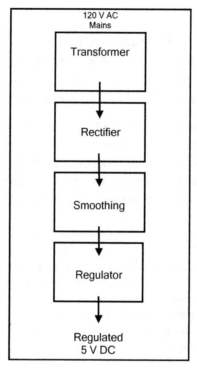

*The process of turning AC to DC power.*

## Transformers

As noted in the preceding section, transformers reduce or increase AC voltage. They do this through the operation of mutual *induction*. A transformer consists of two coils (also called windings) separated by a laminated iron core. AC power is applied to the first coil, creating an electromagnetic field in the core. A secondary current is transferred to the second coil. The secondary current isn't conducted across the core, however; instead, it is induced through electromagnetism.

**DEFINITION**

**Induction** is the production of a voltage by passing a conductor moving through a magnetic field.

Each coil has a certain number of turns, and the ratio between the number of turns of each coil determines the voltage induced in each coil. If we call the voltage in the primary coil $V_P$, the number of turns in the primary coil $N_P$, the voltage in the secondary coil $V_S$, and the number of turns in the secondary coil $N_S$, then the ratio is $V_S/V_P = N_S/N_P$. If the number of turns in the primary coil is more than the number of turns in the secondary coil, the voltage is reduced, or stepped down, across the transformer. A transformer can also "step up" or increase the voltage if the number of turns in the secondary coil is more than the number of turns in the primary coil.

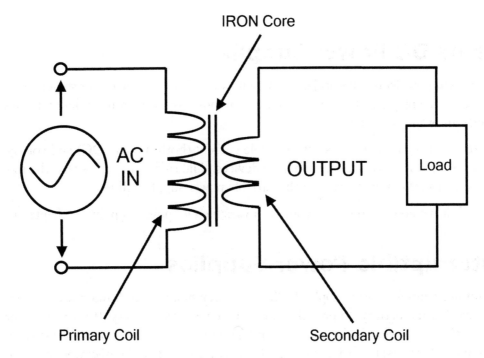

*Diagram of a transformer.*

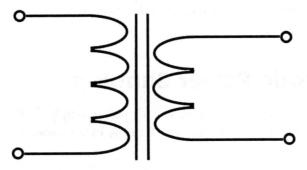

*The symbol for a transformer in a circuit diagram.*

**WATTAGE TO THE WISE**

When a power supply is plugged in to a wall outlet, it is drawing power even when you aren't using the device it powers. AC is still drawn from the outlet and powering the conversion and conditioning of the AC to DC. To conserve energy and save on your electric bill, unplug these power supplies or plug them into a power strip that can be switched off when they are not in use.

# Variable DC Power Supplies

For working on electronics projects, you will want either to buy or construct a variable DC power supply. This is a power supply that connects to the wall outlet and allows you to choose an output DC voltage.

There are also power supplies with variable current (both AC and DC) and voltage; some allow you to test power draws of various devices. The rule of thumb is that the more control you have, the more expensive the variable power supply will be.

You will learn how to construct your own variable DC power supply in Chapter 13.

# Uninterruptible Power Supplies

An uninterruptible power supply (UPS) is a power supply that draws power from two sources (AC and battery) to provide power and charge the battery. If there is an interruption to the AC power, the battery switches over to be the power source. Usually used in computer situations, the battery has charge sufficient to give the operator the opportunity to shut down the computer safely so that there is no data loss or damage to the system.

The difference between a UPS and a backup generator is the near-instantaneous switch between power sources. This is essential in situations where there is a threat of data loss or other damage with power loss.

# Switched-Mode Power Supplies

Primarily used in computers, a switched-mode power supply (SMPS) takes the AC mains power and provides power at different levels to the various voltages needed for the circuits in a computer.

SMPS constantly switches very quickly between on and off, using this rapid switching to manage the various voltages in a computer instead of having specified voltage levels constantly delivered as in so-called "linear" methods.

SMPSs are much more complex than other power supplies, but they are also more efficient. They are able to provide high current for modern computer processing units (CPUs) and can handle a greater range of input and output voltages. An added benefit of SMPSs is that they are less likely to draw power when switched off.

## The Least You Need to Know

- AC mains power and DC from a power supply are the two power sources for electronics.
- Batteries operate without reliance on AC mains power. Electrical energy is produced by a chemical reaction in a voltaic cell. There are two types of batteries: primary, which deplete, and secondary, which are rechargeable.
- Power supplies take electrical energy from a mains power source and condition it to meet the requirements of a device. Most electronics devices need DC at lower voltages than the mains source does.
- Variable DC power supplies are a useful tool for working with projects requiring different output DC voltages.
- Uninterruptible power supplies provide power from one of two sources: from AC mains while power is available and via a rechargeable DC battery when there are power interruptions.
- Switched-mode power supplies provide active power management instead of linear power management. They are used to supply power to personal computers.

## Lab 11.1: Making a Potato Battery

Although building a potato battery is admittedly low-tech, it shows you that the chemical reactions that create electrical current are to be found everywhere around us.

**Materials:**

> Digital multimeter (DMM)
>
> Clip leads
>
> 1 potato
>
> 1 dime
>
> 1 penny

**Instructions:**

1. Push the edge of the dime halfway into the potato.

2. In a spot about 2 inches away from the dime, push the edge of the penny halfway into the potato.

3. Connect the probe leads to your meter: black to COM and red to VΩ.

4. Attach the clip leads to the meter probes.

5. Clip the lead that is attached to the COM or black probe to the penny.

6. Clip the lead that is attached to the VΩ or red probe to the dime.

7. Move the dial to V DC. The DMM should register a very low voltage reading. You have just made a battery.

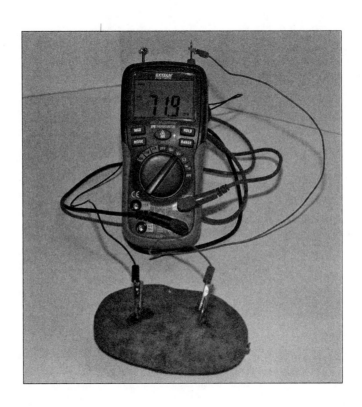

# Getting to Work

Now that you have a grasp of the major electronics components, it's time to learn how to connect them together in a circuit. But to do that, you need to know how to use a soldering iron and solder.

This part also explains the different types of power sources, from potato batteries to more practical power supplies you can use in your projects. By the time you're done with this part, you'll have made your own direct current (DC) power supply.

# Soldering

## In This Chapter

- Connecting parts of a circuit with solder
- Getting acquainted with your soldering iron
- Using flux to prevent oxidation and improve the solder connection
- Working with circuit boards
- Desoldering poorly executed joints

Soldering is an ancient art, dating back to at least ancient Egypt, where soldering was used in creating jewelry. It is the use of a material (solder) to create a joint. For electronics, the process is specifically used to create joints that allow for conductivity of electricity and to protect against water movement through a joint. It is done by applying a melted layer of filler material (solder) to bridge the joint. Unlike welding, which melts the base metals to combine the two components being joined, soldering melts only the solder that forms the joint between components.

Soldering is used in a number of nonelectronic applications, including plumbing, jewelry-making, creating the separation between colored glasses in stained glass, and attaching flashing on roofs. Each of these applications has different requirements with regard to the type of solder and flux used and the temperature needed to make the joint. Some tools and techniques overlap among the various soldering applications, but soldering in electronics has particular requirements to achieve optimum results.

# Solder

Solder is a wire-shaped *alloy* sold in spools. The most commonly used solder in electronics is a 60/40 alloy of tin and lead, which has a melting range of 183°C to 190°C (361°F–374°F).

 **DEFINITION**

An **alloy** is a combination of two metals. Different alloys have different melting ranges.

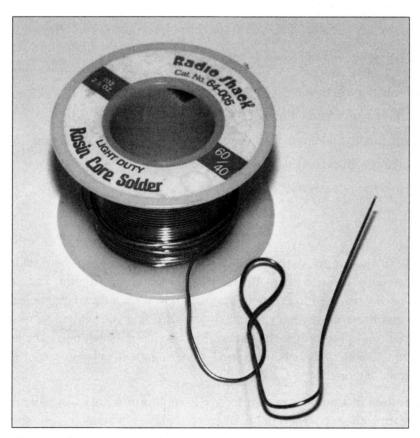

*A spool of solder.*

The 60/40 tin/lead alloy is close to what is called a *eutectic* alloy, a type of alloy that has a single melting point rather than a range. (A true eutectic alloy has a 63/47 tin/lead combination.) The smaller the melting range, the easier it is to solder a joint, as the solder changes from liquid to solid very quickly rather than slowly solidifying as the temperature decreases.

**WATTAGE TO THE WISE**

The metric system is the common language of electronics worldwide. Familiarity with the metric system translates into more efficient use of your time (no need to convert from metric to imperial units) plus fewer errors while working in your shop.

Instead of converting metric units into imperial units (Fahrenheit, ounces, miles), embrace the metric world so the values have meaning to you independently of their imperial conversions. If you use the units of metric measurement regularly, they will become second nature. Sticking with metric temperature units is particularly time saving because the conversion between Celsius (C) and degrees Fahrenheit (F) temperatures involves two steps: readjusting to zero, and then applying the ratio between them. That's a lot of math!

## Lead-Free Solder

Increasingly, governments around the world are prohibiting the use of lead in the manufacture of electronics because of the negative health consequences associated with lead ingestion. Lead can also damage the environment when it is released into the environment in the manufacture or in the recycling of electronics.

Manufacturers are quickly moving to lead-free soldering, and even hobbyists should prepare for the eventual phase-out of lead alloys. Because most lead-free solders are even less eutectic than our 60/40 tin/lead standard, they require better technique to form a good solder joint.

# Flux

In soldering, flux is a compound that prepares a surface base metal so that it can be joined efficiently with solder. It is often said that flux is a cleaning agent to prepare components for soldering, but it doesn't clean dirt and grime. Instead, it reduces oxides to prevent oxidation during the soldering process. When heat is applied to copper, oxidation often occurs, preventing the formation of a good solder joint. Flux prevents the oxidation process.

Flux also assists in wetting, which is the process of reducing the surface tension of the base metal so that the liquid solder can make better contact with it. To get a better sense of wetting, think of a bead of water on a newly waxed car. The water forms a bead because of the greater surface tension of the waxed surface; in other words, the water does not make good contact with the surface. Flux prevents beading of solder, enabling the solder to adhere as a flat droplet instead of an angled bead.

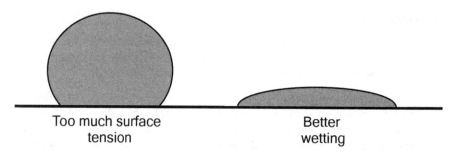

| Too much surface tension | Better wetting |

*Flux improves wetting by decreasing surface tension.*

Solder is often sold already combined with flux, in the form of *rosin*-cored or flux-cored solder. This is often all the flux you need in a basic printed circuit board (PCB) solder. If you are working with surface-mount soldering, you may also want to prepare the surface of the PCB and the component itself with additional liquid flux.

 **DEFINITION**

**Rosin** is a pine tree resin that has been used as flux for many hundreds of years for its ability to reduce friction.

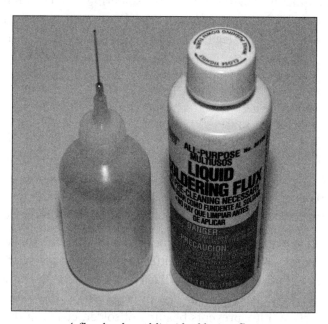

*A flux bottle and liquid soldering flux.*

# Soldering Iron and Tips

The type of soldering iron we recommend (see Chapter 4) has variable power but no temperature control. Generally, if you're just starting out in electronics, you should look for a relatively low-power unit (25–50 W) that can accept a variety of tips and comes equipped with a stand and an on/off indicator light. Our iron also has a place for a sponge for wiping off any solder residue after each use. If yours doesn't come with a sponge, you'll need to buy one and keep it on hand when soldering.

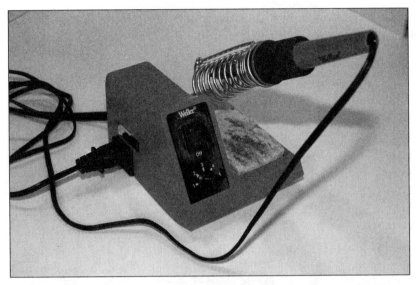

*Soldering iron.*

You will also want at least two different types of tips: one with a flat head much like a screwdriver (this is the one that will come with your iron) and one with a fine point. Tips are usually made of a copper core because of copper's conductivity. The tips are then covered with iron, chrome, and nickel to provide hardness and better high-temperature performance.

*You'll want a flat-head tip (left) and a fine-point tip (right) for your soldering iron.*

Read the manual that comes with your iron before beginning to use it to find out how to properly attach your tips and other advice on operating the iron.

# Heat Sink

A heat sink is used to protect heat-sensitive components from damage during the soldering process. Basically, it is a piece of metal that increases the surface area available for the release of excess heat. As we discussed in Chapter 4, there is a specific electronics tool called a heat sink, but you can also use alligator clips for this purpose.

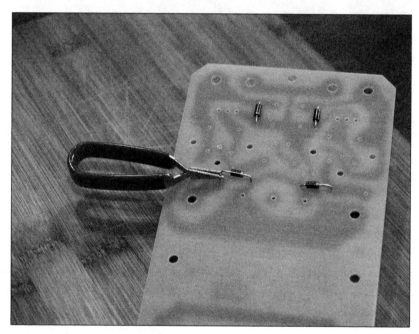

*Using a heat sink.*

# Circuit Boards

The type of circuit board you use to construct your circuits depends on how permanent of a project you are creating. When you're doing quick labs or constructing test circuits and plan to reuse the board, a breadboard is the best choice because it enables you to experiment without making permanent connections.

*A breadboard.*

For more permanent construction, you can work with stripboard or perfboard, two inexpensive boards available from any hobbyist shop. On these boards, you make all of your connections through the copper channels on the board. These boards work with through-hole methods of connecting components; in other words, the leads go through the holes of the board and are soldered to the conductive channels. Both sides of the boards can be used.

## Printed Circuit Boards

Alternatively, you can design and create your own printed circuit boards (PCBs). PCBs use a layer of copper bonded to a substrate that is covered with a nonconductive coating. The desired tracings (or conductive pathways) are etched through to the copper layer. You can print onto blank PCBs or you can order a printed PCB from one of many online printing services.

PCBs are available in both through-hole and surface-mount styles. The connections are made with the copper traces (lines of copper connective) on the board's surface instead of connecting components through holes. The printed areas include solder pads, which are designated solder points for your components. Both sides of the board are available for printing.

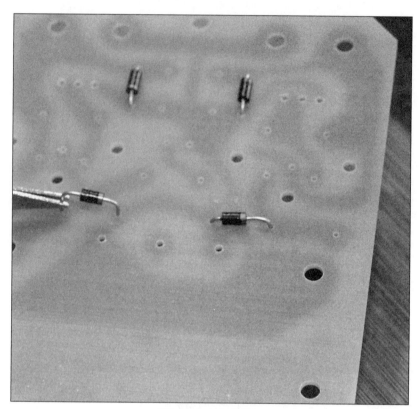

*Through-hole PCB.*

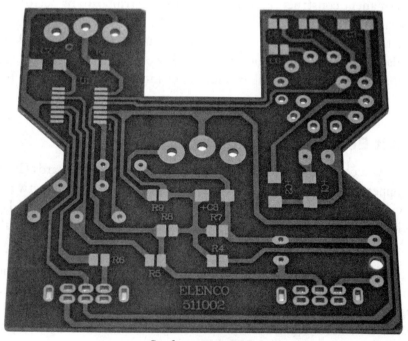

*Surface-mount PCB.*

When you are repairing electronics, you will likely encounter only surface-mount PCBs, so you'll need to become comfortable soldering on these types of boards.

**HIGH VOLTAGE!**

Electronics manufacturers often use multilayer PCBs to connect multiple circuits compactly. Because of the layered construction, it is difficult to access the connections, effectively making most home repairs unfeasible.

# Prepping for Soldering

Before beginning a soldering project, you must first prepare your workspace and materials. In Chapter 5, you learned basic shop safety; take the time to review the key points in that chapter. Make sure you have adequate ventilation so the irritants in the flux and solder are diluted by fresh air. You should have adequate room to work and to organize your equipment and components.

Make sure that the components and the board you are using are clean. Use a brass sponge (with no soap) to remove any waxy or oily substances.

## Place the Items on the Board

Place the items you are soldering together on your board. With a through-hole PCB, bend the lead so it makes a connection. For a surface-mount PCB, carefully place the lead against the exposed copper tracer.

Surface mounts can be more difficult to work with because it's not as easy to keep the components in place while you are soldering them together. One trick is to melt a bit of solder on the solder pad, place the component before the solder cools, heat the component's lead, and then apply more solder. Prepping the areas to be soldered on the PCB with liquid flux prior to soldering can be helpful as well, because the flux helps prevent solder bridges (unintended joints) between part leads that are very close together.

## Prepare Your Solder Gun

Prep your solder gun by following these steps:

1. Make sure your iron's tip is clean and shiny. Know which part of the tip is the proper work area.

2. Prepare your sponge by dampening it (not soaking it until it is dripping wet).

3. Turn your iron on. Follow your manufacturer's instructions for the proper settings for the solder you are using.

4. Wipe the iron's tip on the damp sponge and apply a bit of solder to the tip to tin it. Tinning helps in the even heat transfer from your iron to the solder, so even if your tip is pre-tinned, it's a good idea to re-tin every time you wipe the tip with the sponge. You can also buy tinner, which is a dip-in product that cleans and tins tips.

**HIGH VOLTAGE!**

Irons are hot! Follow these safety tips when soldering:

- Don't touch a hot iron's tip.
- Don't leave a hot iron unattended.
- Don't set a hot iron on anything but its stand.
- If you get burned, follow basic burn first aid.
- Wait until the iron has cooled completely before storing it.
- Know where your fire extinguisher is. Be sure it is rated for electrical fires (Class C) and is up to date.
- Keep the iron's power cord out of the way so you don't risk tripping over it.

# Soldering Technique

A steady hand is always your best tool. Hold your iron like a pen, with a comfortable, relaxed grip. Then follow these steps:

1. If you are soldering a particularly heat-sensitive component (such as a transistor), attach a heat sink to the lead.

2. Heat the connection you want to make, not the solder. Hold the connection with the tip for a few seconds.

3. Apply a little solder and let it flow into a small volcano shape over the connection.

4. Remove the solder, then slowly remove the iron, all while maintaining a steady hand so as not to disrupt the joint.

5. Keep everything still as you inspect the joint. If it looks like it needs more solder, repeat steps 2 through 5.

Don't worry if you make a mistake. You can always remove the solder and reapply it.

Bad joint

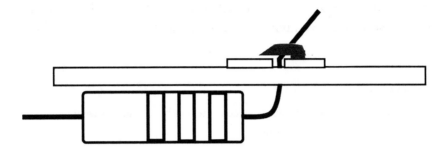

Bad joint

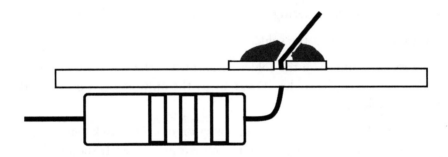

Good joint

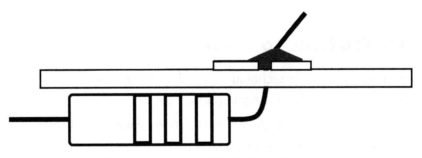

*Examples of good and bad soldered joints.*

# Desoldering

If any solder joints are cracked, incompletely connected, or overlap onto other components or traces, it's best to start over, remove the solder (a process called *desoldering*), and then try again.

**HIGH VOLTAGE!**

Keep in mind that melted solder can cause severe burns. Always wear protective gear and use caution when soldering.

Follow these steps for desoldering:

1. Clean the area around the solder joint. Again, steel wool should do the job nicely.

2. Gather your tools: soldering iron, solder sucker, and soldering wick.

3. Heat the joint to melt the solder.

4. Depress the plunger on the solder sucker; when the solder melts, use the solder sucker to take up the solder. This may be the only step you need to take to remove all of the solder.

5. If there are bits of solder left, heat a portion of solder wick (enough to hold the remaining solder) and place it on the remaining solder. It should suck up the remaining solder.

6. Clean the area with steel wool to remove any remaining rosin or solder bits.

## The Least You Need to Know

- Soldering is an ancient technique of making connections.
- Most electronics solder is a 60/40 tin/lead solder.
- Flux is used to prevent oxidation and to improve the flow of solder.
- There are specialized tools for soldering.
- If the joints are poorly executed, you can desolder the joint and retry the solder.

# Constructing a Power Supply

Chapter **13**

## In This Chapter

- Building a variable direct current (DC) power supply from a kit
- Studying the schematic
- Producing various DC voltages
- Troubleshooting common problems

A variable power supply is an essential tool for any student of electronics. You can buy one already assembled, but this chapter teaches you how to construct one using a kit. Doing it yourself gives you hands-on experience producing an electronics project using the tools you've gathered and provides you with an opportunity to apply some of the concepts you've learned up to this point.

## Power Supply Kit and Construction

The power supply kit described in this chapter is from Jameco Electronics (www.jameco.com); the part number is 20626(JE215). The kit comes with all necessary parts and a circuit board. Once it's constructed, you'll have a general-use power supply that you can use in the projects and labs covered in the rest of this book. You can purchase another variable DC power supply pre-assembled, but by building your own (either with this kit or another) you can practice your soldering skills and gain some hands-on experience with the components in a power supply. Basic kits generally range from $25 to $40.

**WATTAGE TO THE WISE**

Don't worry if you have a different model or brand of power supply kit. The same principles apply to all power supplies, so you should be able to follow along with this chapter using any power supply.

Using the set of instructions that comes with the kit, assemble the power supply. If you accidentally break a part, you can order replacements (see the instructions for details). But if you follow the instructions and use a heat sink on the diodes, you shouldn't have any problems.

**WATTAGE TO THE WISE**

The schematic (circuit diagram) and instructions that comes with your kit has lots of symbols and abbreviations. Each of the symbols represents a different component, which we include in this chapter. Learning to use a schematic takes patience but no special talent.

*The top of printed circuit board (PCB) diodes and the heat sink. Note the placement of the diodes and how to apply the heat sink before soldering.*

*Bottom of the PCB. Note the leads and the copper traces and the first soldered joints in the upper-left side of the PCB.*

*Finished top view. Your supply will look like this when completed.*

*Finished bottom view. Don't forget to place the feet to raise the power supply's conductive traces. Even with the feet applied, we recommend only using your power supply when it is placed on a nonconductive surface.*

**HIGH VOLTAGE!**

This power supply uses household alternating current (AC) for its power source. Household AC mains in the United States deliver current at 120 V; worldwide, the range is 100 V to 250 V. Serious injury or even death can result from electric shock at these voltages. You should not be terrified of working around AC, but you need to be alert and double-check everything you do.

This supply produces adjustable positive and negative voltages ranging from 1.2 VDC (volts of DC) to 15 VDC and power outputs of ±5 VDC at 500 mA, ±10 VDC at 750 mA, ±12 VDC at 500 mA, and ±15 VDC at 175 mA. All of the digital labs later in the book require ±5 VDC at 500 mA.

# Safety First

Turn over the power supply. Do you see those traces—the copper lines in the PCB? Each one is conductive. To keep things safe, use a large wooden cutting board as a work surface. Be very careful about where you place your hands and test probes when the power is on.

**HIGH VOLTAGE!**

These traces can carry live current; if they come into contact with another conductive surface, such as your skin or metal, they can cause an electrical shock.

We also recommend using a power strip that has an on/off switch as this power supply does not have its own power switch—the only way to turn it off is to unplug it. Using a power strip will save wear and tear on the power cord.

Another good practice is to use test probes with clips that will hold the probe in place so you don't have to hold it in place by hand. This way you can place your test probes where they need to be with the power off and step back before turning the power on. This process can help you build confidence in what you are doing as well. Remember, people work safely with AC power every day.

## Powering On

After you've assembled the kit, it's time to start exploring what you can do with your new power supply.

First, plug the power supply in to a power strip with the switch turned off. Then turn the power strip on.

You should see LED1 light up; if not, there is a problem. Refer to the instructions that came with the power supply for troubleshooting tips. In addition, you can check for the following for commonly encountered problems:

- Is the LED properly installed in the proper position?
- Is the power cord connected correctly?
- Check for the proper placement of C1 and C2 the + and –.
- Check for 18.5 VDC across C1. Attach the black probe to the – side and the red probe to the + side, and set your DMM to VDC. Be sure to also check the same across C2.

**WATTAGE TO THE WISE**

Troubleshooting is an art. Generally, the more experience you get, the better trouble-shooter you will be. Thankfully, for the beginner, there are a few approaches you can take when you are first starting out. First, check for the basics, such as whether the plug is plugged in all of the way. Sometimes it is the easy stuff. Second, keep an open mind. A little imagination goes a long way. Third, work step by step. Troubleshooting requires deliberate elimination of any potential problem. Remember that the shortest path between two points is a straight line; if you take a scattershot route to a solution, you might find yourself getting lost—and frustrated—along the way.

# Getting Acquainted with Your Power Supply

For the following procedures, you will need a digital multimeter (DMM) and, if you have one, an oscilloscope. If you don't have an oscilloscope, don't worry; we provide rough diagrams of what you would see on a scope.

First, let's look at the schematic of the assembled power supply that's included in the instructions.

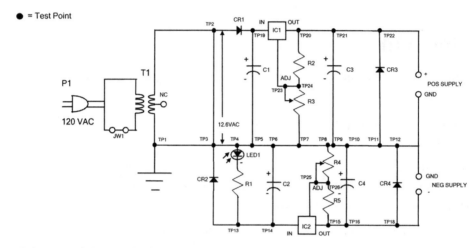

*Schematic of the assembled power supply. Note the use of the circuit diagram symbols you have learned so far.*

(Courtesy of Jameco)

The filled dots are test points. They are labeled TP1 through TP26.

## Creating a Positive DC Wave

To show the function of your power supply, let's turn it on and generate a positive DC voltage signal. By using the different test points you can see many of the components you have learned about so far in action:

1. Using your DMM, place the black probe on TP3 and the red probe on TP2.

2. Set your DMM to Vac (volts AC).

3. Turn on the power. You should have a reading of around 12.6 Vac. This is the effect of the power supply's transformer stepping down the 120 Vac from the mains to 12.6 Vac, as described in Chapter 12.

4. If you have an oscilloscope, place the probe on TP4 and place the ground clip on the circuit ground (labeled GND on the schematic). You should see a positive pulsing waveform (also known as a rippling DC wave).

5. Ripples or pulses, which are variations in the current, occur when the current has not yet been smoothed by the capacitors. The power supply at this point has removed the negative portion of the AC, but it is not yet ready to be used as a DC power source.

6. Even though this current needs to be further smoothed by the power supply, it does provide power to the LED, letting us know that the power is on.

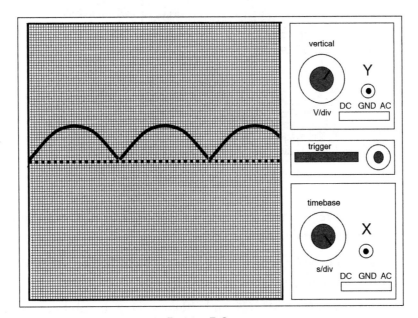

*Positive DC wave.*

## Seeing Caps and Voltage Regulators in Action

Now let's look at your power supply's capacitor in action. In this case, it has a slower discharge time than the frequency of the input signal, so it gets charged more quickly than it can discharge. This is a good thing for a DC power source, as it will smooth out most of the ripple you saw in the previous test.

Place your scope probe on TP19 or TP14 and the ground clip on the POS SUPPLY ground.

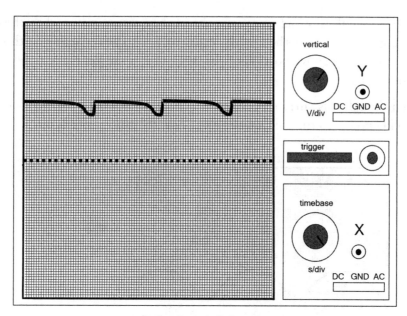

*Reduced ripple DC wave.*

IC1 and IC2 are integrated circuits (IC) that act as voltage regulators and are used to provide a constant voltage and current.

The voltage is adjusted by voltage dividers, which are identified on the schematic as R2 and R3 (on the positive side) and R4 and R5 (on the negative side).

The capacitors and diodes C3, CR3, C4, and CR4 are there to stabilize the output when adjusting R3 and R4 and to protect IC1 and IC2 from feedback on the outputs or an accidental short on a circuit connected to the POS or NEG SUPPLY, as this could destroy the ICs.

CR3 and CR4 also provide a discharge path for C2 and C1 when you shut the power off. As you may recall from Chapter 8, capacitors can hold a charge even after they are powered off.

# Using Your Variable DC Power Supply

Now you have a variable DC power supply that you can use for all of your electronic projects. Let's take a look at the output and how to make adjustments:

1. With the power to the supply turned off, connect your DMM to the power supply. The red lead to the +DC terminal and the black lead to the ground (labeled GND in the next illustration).

2. Attach the black probe to the COM on the meter and attach the probe end to the GND post of the power supply.

3. Attach the red probe to the VΩ on the meter and attach the probe end to the + post of the supply.

4. Set the meter to read DC voltage.

5. Turn the power on and see what output you have.

Try to adjust the output voltage. You can adjust this by using the settings knob of R3. You should see the output voltage changing. If not, check the following:

• Are either IC1 or IC2 installed backward? If yes, they will need to be replaced, as they were most likely destroyed. You will need to reorder new ones using the part numbers provided in the instructions.

• If the output stays high, check for proper solder on R3 and R4.

• If the output is not constant or is hard to adjust, check proper placement of C3, C4, CR3, and CR4.

The other settings knob, R4, changes the negative output voltage.

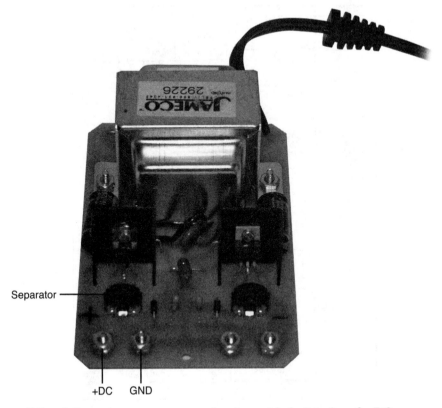

*+DC and Ground on your power supply, along with the R4 adjust for DC output.*

**WATTAGE TO THE WISE**

The projects you will be completing later in the book require a 5 V_DC setting. To change your power supply to that output, use the knob of R3. After adjusting the output, do one last DMM check to confirm that it is set correctly.

# It's Time for Some Comic Relief

You've now been introduced to all of the major players in electronics. You know how electricity flows through a circuit; you know how to stop it, to amplify it, and to direct it. You have seen lots of symbols and, although you might need to go back and look them up again, you understand that a schematic contains all the instructions you need to construct a circuit.

Almost every discipline has its own vocabulary and symbols that are seemingly indecipherable by the untrained. It becomes like a secret handshake, known only by those who have put in the time and study. Congratulations on your initiation into the fold. You are now among the few who can look at a squiggly line on a drawing and see a resistor.

Before jumping to the next part and tackling digital theory, here is a bit of humor provided courtesy of the webcomic XKCD. You've earned a laugh.

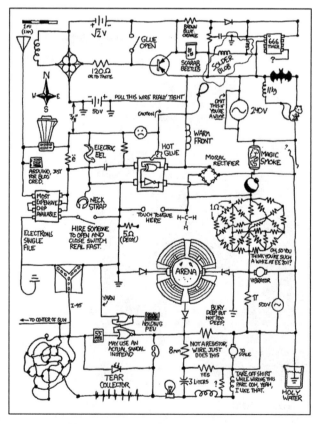

*I just caught myself idly trying to work out what the resistor mass would actually be, and realized I had self-nerd-sniped.*

## The Least You Need to Know

- Follow the instructions that come with your power supply kit to assemble the power supply.
- Plug the power supply into a power strip or a source that has an on/off switch.
- The power supply you made produces adjustable positive and negative voltages of 1.2 $V_{DC}$ to 15 $V_{DC}$ and power output of 5 $V_{DC}$ at 500 mA, 10 $V_{DC}$ at 750 mA, 12 $V_{DC}$ at 500 mA, and 15 $V_{DC}$ at 175 mA.
- A power supply does more than simply convert AC to DC. Capacitors and voltage regulators are used to remove ripple to provide steady DC voltage.

# Going Digital

**Part**

# 5

Transistors were quite a leap in the history of electronics, but another equally revolutionary development was the application of binary values to the on/off (well, really high- and really low-voltage) states, and the idea of representing questions in logic to electronic circuits. The following chapters explain how digital electronics work and how you can put them to work for you.

You will learn about integrated circuits, which are simply multiple circuits miniaturized and combined on a single chip. You will next learn how memory stores data as well as the instructions for running digital devices. You will be able to understand that writing a program is speaking to a machine in its own language.

Microcontrollers are amazing digital tools. They are small computers designed to work in embedded devices, including your own electronic projects. They have a streamlined construction and are easy to customize.

# Digital Theory

## In This Chapter

- Using binary digits in electronics
- Distinguishing between analog and digital electronics
- Representing logical operations with truth tables
- Using logic gates to control circuits

The idea of linking the binary digits 1 and 0 to high-voltage and low-voltage levels respectively made possible revolutionary changes in how we use electronics. Great strides had already been made in electronics prior to this digital revolution, especially in the field of communications and the launch of radio and television. But once digital concepts were overlaid on electronics theory, the pace of change in technology accelerated to warp speed.

## The Idea Behind Digital Concepts

In 1937, a Massachusetts Institute of Technology (MIT) graduate student named Claude Shannon wrote "A Symbolic Analysis of Relay and Switching Circuits," which has been called one of the most important Master's theses of the century. In the paper, Shannon proposed the use of Boolean algebra's two-position analysis in creating digital circuits. The world of electronics was from then on destined to change, even if it took a few decades for the idea to reach its full potential.

## Analog vs. Digital

An analog signal is a continuous wave in both amplitude and in time. A digital signal is composed of individual, identifiable steps. A wave in the ocean is an analog signal; the tick-tick-tick of the second hand on a clock is digital. The following figure shows a sine wave in analog form and digital form. The digital wave approximates the analog wave by representing points along the waveform.

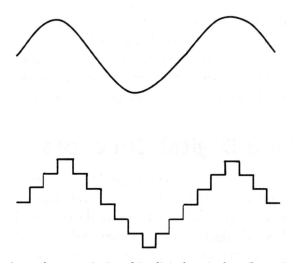

*An analog wave (top) and its digital equivalent (bottom).*

There are pros and cons to working with each signal type. Because an analog wave is a continuous wave, it has more information. However, because it has more information, it needs more processing and storage space. A digital signal can be much more compact than its analog counterpart, which means it requires fewer resources to transmit, receive, or store. A digital signal is also less likely to be affected by *noise*.

> **DEFINITION**
>
> In electronics, **noise** is any unwanted interference with a signal. Noise sources can be natural, due to the natural interaction of electromagnetic fields, or man-made—from motors, fluorescent lights, radio or radar transmissions, wireless signals, and numerous other sources.

One of the primary benefits of digital waves for electronics is that each discrete bit of information can be expressed numerically using *Boolean algebra*. The states of "on" and "off" or "high voltage" and "low voltage" can be represented by 1 or 0, which in turn can be easily transmitted using switches, diodes, and transistors in circuits.

> **DEFINITION**
>
> **Boolean algebra** is an approach to the study of numbers that is based on logic.
>
> **Logic** is a branch of philosophy that was developed by the ancient Greeks, including Aristotle. It starts with the basic premise that an answer is either true or false. Aristotle proposed that there are laws of logic regarding a bi-valued reality in which statements are either true or false: X = X, X ≠ Y, not–X = Y, and X = not–Y. English mathematician George Boole expanded upon these classical views of logic and developed a system of algebra based on them.
>
> Boolean numbers are not the same as real numbers. Instead, Boolean numbers represent a decision. They ask: Is it 1? If yes, 1. If no, 0. Each decision has the possibility of producing just two outcomes—1 or 0.

# Truth Tables

The most common questions asked about a relationship of numbers in Boolean logic can be expressed by the following logical operations:

- NOT: negation
- AND: conjunction
- OR: inclusion
- NOR: neither/nor
- NAND: not both
- XOR: exclusive
- XNOR: equality

To represent these concepts, we can use something called a *truth table*. One value, called an *operand*, goes across the top of the table horizontally and the second operand goes down the table vertically. For each question or logical operation, the various answers populate the box.

An **operand** is a quantity that has a mathematical or logical operation performed on it.

A **truth table** is used in Boolean logic to give the results for the possible inputs and outputs. The columns represent one side of the logical decision and the rows represent the other side of the logical decision. The result of each combination is depicted in the box where the column and row intersect.

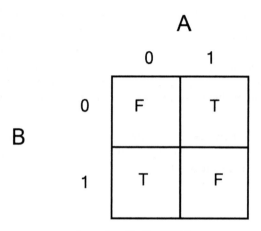

*A truth table for NOT.*

In this table, $0 \neq 1$ is true (because 0 is NOT equal to 1), whereas $0 \neq 0$ is false. Let's now represent true with 1 and false with 0, as shown in the following figure.

A

|   | 0 | 1 |
|---|---|---|
| 0 | 0 | 1 |
| 1 | 1 | 0 |

B

*Truth table for NOT with binary substitution.*

This table does not represent multiplication, division, or any other real mathematical operation. Instead, it is the outputs of the question, "Is A not equal to B?" The answer can be true or false, and we are using the binary digits 0 and 1 to represent both the operands and the answers.

Truth tables can be created for each of the logical operations you just learned. Let's look at AND. The operation AND is true only if both operands are true—in other words, only when both operands are 1.

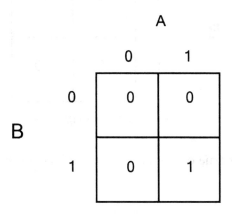

*Truth table for AND.*

The operation OR is false only if both operands are false—that is, only when both operands are 0.

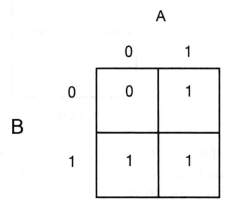

*Truth table for OR.*

The operation NOR is true only if both operands are false—only when both operands are 0.

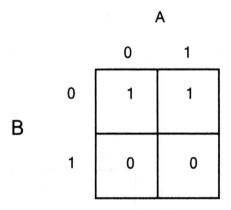

*Truth table for NOR.*

The operation NAND is true only if at least one of its operands is false. One operand must be 0 and one must be 1.

A

|   | 0 | 1 |
|---|---|---|
| 0 | 1 | 1 |
| 1 | 0 | 0 |

B

*Truth table for NAND.*

Two common operations of Boolean logic that we use in electronics require two inputs: XOR or "exclusive-OR" and XNOR or "exclusive-NOR." These operations compare the two inputs and ask a question about them.

Here is the truth table for XOR. Looking at the two inputs, the output is true if one (and only one) of the inputs is true.

| | | |
|---|---|---|
| 0 | 0 | 0 |
| 0 | 1 | 1 |
| 1 | 0 | 1 |
| 1 | 1 | 1 |

*Truth table for XOR.*

The XNOR gate is the inverse of XOR. For XNOR, looking at the two inputs, the output produces a value of true if and only if both operands are false or both operands are true.

| | | |
|---|---|---|
| 0 | 0 | 1 |
| 0 | 1 | 0 |
| 1 | 0 | 0 |
| 1 | 1 | 1 |

*Truth table for XNOR.*

# Binary Numbers

Binary numbers are distinct from Boolean numbers. Binary numbers use the same two digits as Boolean numbers, but binary numbers are real numbers that represent values beyond 0 and 1. (See Appendix D for details on working with binary numbers.) A binary system can represent numbers from the base 10 system (our familiar numbering system with digits from 0 to 9) with a string of binary zeros and ones. Binary number representations of decimal numbers or binary-coded decimals (BCDs) can be sent as individual digits (in series) or as a group of *bits* (in parallel). The usual arrangement is in a *byte*, which is eight bits.

# Application to Electronics

So why do you need to know about binary numbers and Boolean operations? Because of the real-world application of these concepts to electronics. You've already learned that, using diodes and transistors, you can create circuits that can pass through no (or relatively low-) voltage or relatively higher-voltage signals.

If you go back to the idea of a computer as something that performs computations, you can see that creating a series of questions or logical operations could get you to a result. Of course, the number of operations required to perform even simple tasks would require lots of these operations and, therefore, lots of circuits.

# Logic Gates

Each of the circuits that do these computations contains gates. In digital or logic circuits, the gates are called *logic gates*. Each gate gives one output but can have multiple inputs.

NAND gates and NOR gates are known as *universal gates* because given enough combinations, they can mimic the function of any other logic gate. A NOT gate is also called an *inverter* because it inverts one signal into the reverse.

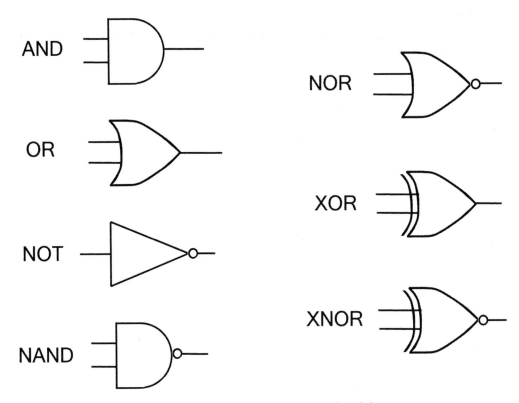

*Symbols for various logic gates on a circuit board diagram.*

**WATTAGE TO THE WISE**

When you begin designing your own circuits or doing your own programming, you need to think like a circuit or a computer. It is important to map out the processes step by step. Designers and programmers often use flowcharts to represent the inputs and the decisions at each step. As you start out in electronics, get in the habit of planning things all out. You might be going high-tech, but a pencil and paper can still be your most useful tools!

## The Least You Need to Know

- Claude Shannon revolutionized electronics by combining binary concepts with voltage levels.

- An analog signal is a continuous wave; a digital signal is composed of individual, discrete steps.

- Boolean algebra is an expansion on classical logic theory composed of statements that are true in a bi-valued world. The answers to each logical question can be depicted in truth tables.

- Using Boolean logic, you can create electronic circuits that can perform logical operations.

## Lab 14.1: AND Gate

To see how a digital logic gate works in action, let's construct an AND gate.

**Materials:**

> 1 9 V battery
>
> 2 single-pole, single-throw (SPST) switches
>
> Jumper wire
>
> 1 flashlight bulb

**Instructions:**

1. Connect the jumper wires as shown in the diagram to the two switches, the lightbulb, and the battery.

2. Turn on SW1 to turn the lightbulb ON. off.

3. Turn on SW2 to turn the lightbulb ON. on.

   The circuit is constructed so that if SW1 = ON AND SW2 = ON, then TRUE. The true answer is represented by high voltage so it turns the light ON.

4. Now let's generate a FALSE output by turning off SW1. This yields a FALSE (or low voltage) because with SW1 = OFF, it is not true that both are on. The result would be the same if both were turned off, as they both need to be ON to be TRUE.

# AND Gate from Switches

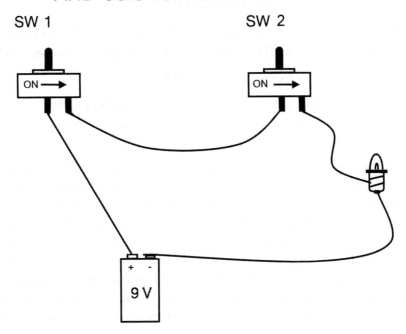

1 & 2 on light is on
1 or 2 on light is off

## Lab 14.2: OR Gate

Let's try another gate, the OR gate. In this case, for the answer to be TRUE, either SW1 OR SW2 needs to be ON.

Use the same materials as in the preceding lab.

1. Connect the jumper wires as shown in the diagram to the two switches, the lightbulb, and the battery.

2. Turn on SW1 to turn the lamp ON.

3. Turn off SW1 to turn the lamp OFF. The lightbulb will light because if we ask the question: SW1 = ON OR SW2 = OFF, the answer is TRUE.

4. Turn on SW2 to turn the lamp OFF.

5. Turn off SW2 to turn the lamp ON. Again, the lightbulb will light because if we ask the question: SW1 = OFF OR SW2 = ON, the answer is TRUE.

The combination of Boolean algebra, the binary system, and the ability to create high- and low-value voltage values with electronic components is one of the most revolutionary concepts in human history. Because of this idea, and all of the steps required to invent the devices that make it all possible, we can use electrical signals to solve the most complicated calculations, to render digital images in timeframes almost too small to measure, to connect people from across the planet and into space. While it's easy to get overwhelmed with the potential of electronic decision-making, it's important that you understand that each thing you do with a computer requires many, many individual logical decisions being made one at a time.

# OR Gate from Switches

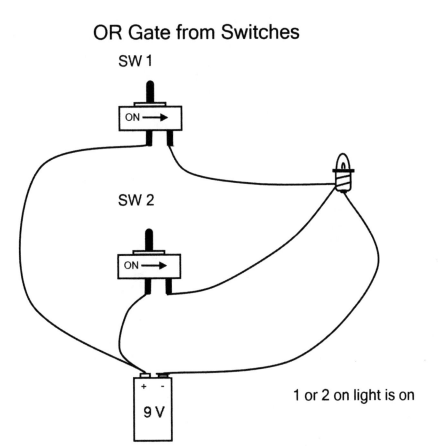

SW 1

SW 2

1 or 2 on light is on

# Integrated Circuits

## In This Chapter

- Integrated circuits (ICs)
- Analog ICs
- Digital ICs
- Mixed signal circuits

An integrated circuit (IC) is a miniaturized circuit that rests on a semiconductor base, also known as a chip. Modern ICs can contain many millions of transistors all in a compact package. ICs can be analog, digital, or a mix of analog and digital signals.

ICs are manufactured on a mass scale, so we won't address their construction, but they are a part of almost any modern electronic device. There were ICs in your power supply project (see Chapter 13) and they will be on most other project shopping lists as you work in electronics.

Each of the ICs discussed in this chapter have standard functions and are common tools in the electronics toolbox. Other ICs are designed for specific applications.

## Analog ICs

Analog ICs work with varying levels of voltages, not simply high or low voltage. They are used in sensors, timers, and power management and as amplifiers. One example is an operational amplifier (op-amp) circuit. Op-amps take input voltages from two terminals and can output a voltage many hundreds of times higher than the voltage that is input. Many high-end audio systems make use of op-amps.

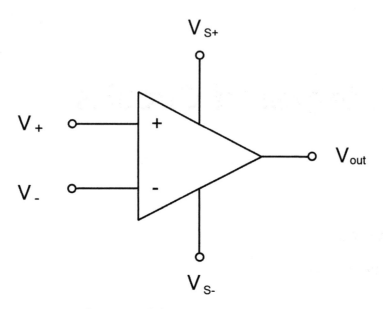

*Op-amp symbol on a circuit board diagram.*

# Digital ICs

Digital ICs, also called *logic ICs*, range from very simple circuits that process a single input from a light sensor to chips that can perform millions of logic operations. Digital circuits perform one of two types of functions: decision-making using logic gates or memory storage.

**WATTAGE TO THE WISE**

You can compare the computing and storage of ICs to the same ability in the human brain. Our brains process data (sensing, processing, decision-making), giving an output and then storing the whole matter into our memory to recall for future use.

Engineers are working toward achieving an artificial re-creation of these abilities in the field of artificial intelligence. Neurobiologists are also analyzing just how our brains do what they do. Imagine the possibilities if scientists are able to create electronic models of human decision-making.

Logic ICs are classified as either Active-Low or Active-High. The IC functions only when a certain voltage level—either high or low—is applied. In a complementary metal-oxide semiconductor (CMOS) with a 5 V power supply, high voltage (or binary value 1) is voltage in the 3.5 V to 5 V range. Low voltage (or binary value 0) is in the 0 V to 1.5 V range.

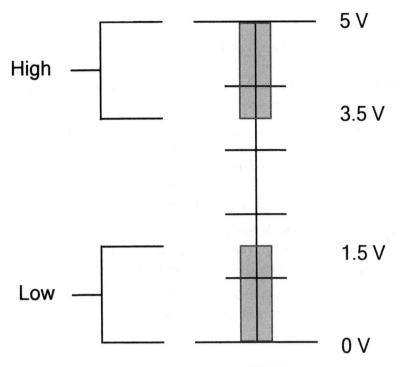

*High or low voltage in a CMOS IC.*

Whether an IC is Active-Low or Active-High is indicated in one of the following two ways in a circuit diagram.

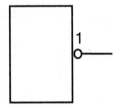

*An Active-Low IC symbol on a circuit board diagram.*

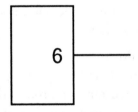

*An Active-High IC symbol on a circuit board diagram.*

The following sections describe various types of digital ICs.

## Flip-Flops

As you've seen so far, a logic gate is designed to produce just one output. It's possible, however, to arrange a circuit that will have the output feedback to the input; that is, the gate constantly feeds its output value back around to the gate as an input value. These types of gates are called *multivibrators*.

One use of a multivibrator is a *flip-flop*. A flip-flop is used to store a single bit in either a high or low state. It can be created by arranging the connections so that the output provides feedback that keeps it in that state of either high or low. This is essential to the concept of storing memory (see Chapter 16 for more on memory). A flip-flop has two outputs, labeled Q and not-Q. The symbol for not-Q is a Q with a horizontal line over it.

The primary output, Q, holds the logic state of the flip-flop. The output at not-QQ is a complementary output, so it holds the opposite value. If Q is 1, not-Q is 0. When Q is 1, the flip-flop is in set mode; when Q is 0, the flip-flop is in reset mode.

The most common flip-flop used in logic circuits is the D flip-flop, where *D* stands for delay. It relies on a pulse circuit to set or reset the output.

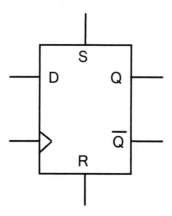

*A D flip-flop circuit diagram symbol; note the symbol for the not-Q output.*

Four D flip-flops can be connected to a clock input (a clock in this case is a timing circuit—a specialized IC—that provides pulses to a system to synchronize operations) to form a storage register, which is a form of memory. Four bits (a nibble) of binary values can be stored at each pulse of the clock input.

Let's go through that again. We don't normally think of electronic signals as being stored. The current travels through a circuit, and then the signal is lost. The memory register uses the feedback to the input to constantly reprocess the output level. This is to keep the memory register showing the same output value in a steady state so it can "remember" the value.

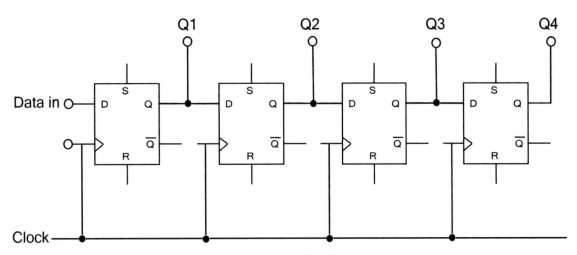

*A storage register is composed of four D flip-flops connected to a clock input.*

## 555 and 556 Timers

With most electronic devices some sort of timing mechanism is needed to enable the various processes within the device to work in sync. When you are working with linked devices such as phones, networked computers, even televisions and radios, the interaction of these devices requires synchronized signals. The more sophisticated the operation, the more precise the timing must be.

There are many methods of providing a clock or timing input, some involving the oscillation of different crystals or nuclear materials, but in most electronic devices a specialized IC known as a timer is used. Timers rely on the frequency of an outside supplied voltage to create a signal that is fed back to the system using digital logic. Two common logic ICs are the 555 and 556 timers. Each has three operating modes that can perform many functions. A 555 IC has 8 pins (leads or connectors) and the 556 has 14 pins.

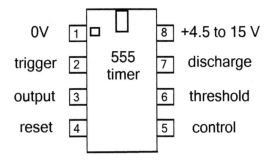

*The symbol for a 555 timer on a circuit board diagram.*

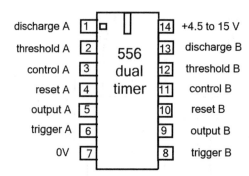

*The symbol for a 556 timer on a circuit board diagram.*

These timers have three distinct modes. In astable mode, the 555/556 pulses a continuous square wave at the frequency you set. It is called *astable* because the output is constantly alternating between two states. Some uses include any sort of pulsing circuit, such as flashing light-emitting diodes (LEDs) or clock pulses.

In bistable mode, the output stays at one state until a triggering input switches it to the other state. This can operate a switching circuit or a flip-flop.

Operating in the monostable mode, the output remains low until a trigger event, and then a single pulse of high voltage is output. This can be used to overcome the electrical noise from the bounce of a switch (the energy created by the mechanical operation of a switch) by sending a reverse pulse, or in any application where a triggering event is needed.

## Counters

ICs can act as counters. Counters are used to count the number of pulses to then trigger another operation or to provide information back into the system. There are two main types of IC counters: ripple and synchronous. A ripple counter counts through a series of flip-flops. Each output state cascades to the next. A synchronous counter uses a single clock pulse.

Ripple counters count on the falling edge of each square wave, and synchronous counters count on the rising edge of each square wave. Many basic counters use *binary-coded decimal (BCD)*.

**DEFINITION**

**Binary-coded decimal (BCD)** is a hybrid of both the decimal and binary system used. Instead of providing a true binary representation of a digital number, the system gives each of the binary equivalents of the individual digits in binary form. For example, the decimal number of 142 in true binary would be 010001110. In BCD, it is 0001, 0100, 0010, to represent the digits 1, 4, and 2. The leading zeros are given because each bit in the nibble would need to be filled. In electronics, each of these significant digits are labeled D, C, B, and A.

## Encoders and Decoders

Encoders are ICs that convert multiple inputs into a single output. An example is your computer keyboard. A standard keyboard has more than 100 characters that could be input into your computer. Instead of each key having a direct input into your computer's central processing unit, an encoder transmits the character name in BCD, binary, or hexadecimal via the encoder chip's single output.

Each value represents a character in ASCII (American Society for Code Information Interchange) or the more recent Unicode standards (UTF-8 is the current Unicode standard). These systems are standards for communicating on the web and via e-mail. UTF-8 encodes each character in one to four octets (8-bit bytes). The first 128 characters of the Unicode character set (which correspond directly to ASCII) use a single octet with the same binary value as in ASCII. The first 32 characters in ASCII (and UTF-8) are nonprinting control characters such as ESC and DEL.

The following ASCII chart gives the binary, octet, decimal, hexadecimal, and keyboard character values.

### ASCII Values

| Binary | Octet | Decimal | Hexadecimal | Character |
|--------|-------|---------|-------------|-----------|
| 100 0000 | 100 | 64 | 40 | @ |
| 100 0001 | 101 | 65 | 41 | A |
| 100 0010 | 102 | 66 | 42 | B |
| 100 0011 | 103 | 67 | 43 | C |
| 100 0100 | 104 | 68 | 44 | D |
| 100 0101 | 105 | 69 | 45 | E |
| 101 0111 | 127 | 87 | 57 | W |
| 101 1000 | 130 | 88 | 58 | X |
| 101 1001 | 131 | 89 | 59 | Y |
| 101 1010 | 132 | 90 | 5A | Z |
| 101 1011 | 133 | 91 | 5B | [ |
| 101 1100 | 134 | 92 | 5C | \ |
| 101 1101 | 135 | 93 | 5D | ] |
| 101 1110 | 136 | 94 | 5E | ^ |
| 101 1111 | 137 | 95 | 5F | _ |

*Decoders* perform the opposite action of encoders. They take an input and create multiple outputs. Decoders are integral to display circuits, memory addressing (the location where computer data is stored, more on memory in Chapter 16), and code translation circuits.

# Mixed Signal ICs

Mixed signal integrated circuits have both analog and digital components. The analog segments can work with power, radio, or sound signals and integrate them with digital controls. Most of these circuits are analog-to-digital converters (ADCs) or digital-to-analog converters (DACs).

An important concept to understand with ADCs or DACs is sampling. Because an analog signal is continuous rather than the stepped signal of a digital signal, to convert an analog signal to a digital signal it is necessary to sample the wave. The more samples taken of the analog signal, the more accurate the representation of the signal in digital form will be. This concept is easily understood if you listen to the difference in the quality of the sound produced by a voice phone call (sampled at frequencies ranging from 8,000 to 16,000 Hz) versus that of an audio CD (which has a sampling rate of 44,100 Hz). The CD has a higher sampling rate than a voice phone call, so there is a much fuller sound.

**HIGH VOLTAGE!**

It is good practice to turn your power supply off after you complete each lab to avoid any accidental electric shocks. As an extra check, be sure it is off before starting any project.

## The Least You Need to Know

- An integrated circuit (IC) is a miniaturized circuit that rests on a silicon or other semi-conductor base.
- An operational amplifier is an example of an analog IC.
- Digital ICs, also called logic ICs, are classified as either Active-Low or Active-High. Some examples of logic ICs are flip-flops, encoders, decoders, and timers.
- Mixed signal circuits can be digital-to-analog converters (DACs) or analog-to-digital converters (ADCs).

# Lab 15.1: Building a Decoder Circuit, Part 1

In this lab you will use a DIP switch to send a binary signal to a decoder. You will see how when you change the switch settings, or the inputs, it changes the output on the display.

**Materials:**

Breadboard with 5 VDC power supply

4 or 8 dual-inline-package (DIP) switches

1 7448 decoder

7 270 Ω resistors

1 common cathode 7-segment LED display

Jumper wire

In this lab you will learn to use a schematic instead of a picture showing the components on the breadboard. You should recognize many of the symbols from previous labs like the 270 Ω resistors between the decoder and the LEDS, but in this case the LEDs are all in the 7-segment display. Schematics may seem complicated to read at first, but they give you all of the information you need. They are also much less cluttered than the same circuit shown as a photograph. Compare the schematic below with the photograph of the same circuit in the photograph following the schematic.

Only use 4
switches

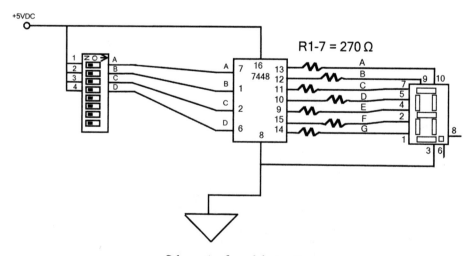

*Schematic of our lab circuit.*

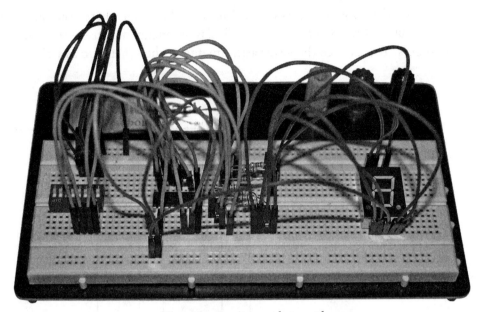

*The same circuit as a photograph.*

Before we leave you to your own devices with a schematic, let's take the time to learn a bit more about using a breadboard. In the lab in Chapter 6 you learned about the power rails of the breadboard (the outer two horizontal rows on the top and bottom) and that the two larger work areas in the middle are where you should build your circuits. Note: when we talk about rows and columns we are looking at the breadboard aligned so that the breadboard is wider rather than taller. So how do you get the power to the work area?

From the power rail row (horizontal), use a jumper wire to connect to a cell in the work area. When you add power to the work area, all of the cells in the column (vertical) that you place the jumper wire in will now have power. Keep this in mind as you construct your circuit. Leave enough room to fit all of your components along the rows so you don't cross columns that are connected to an earlier powered column or component lead.

Be sure to connect your circuit to the power rail that leads either to the negative end of your battery or the ground lead of your power supply.

On the 7448 decoder you will need to be able to find pin 1. This is where you will start making your connections. If you look at the following figure you see that pin 1 is on the top left of an IC on the side where this is a small notch. Pin 2 is underneath it going down vertically. The 7448 decoder has 16 pins—8 on each side forming a U ending at pin 16 on the opposite side across from pin 1. Finding pin 1 on the 7-segment display will depend on which 7-segment display you purchase.

The data sheet for your decoder will be available from where you purchased it or searchable online. Some 7-segment display segments may use letters instead of numbers. You can see the letter assignments in the schematic as well.

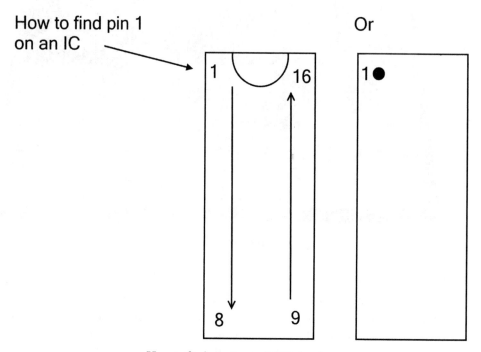

*How to find pin 1 on a 7448 decoder.*

Also note the ground symbol (GND). This is the return path to the ground terminal on your power supply.

**Instructions:**

1. Before assembling your circuit, make sure your power supply is off.

2. Use the breadboard to connect the circuits as shown in the diagram. Make sure you connect the ON side of the DIP switch to the correct pins of the 7448 decoder. The resistors are labeled R1-7.

3. Connect your power supply to your breadboard. See the following figure to identify the +DC and GND terminals. The wire from the +DC terminal should connect to the top row of your breadboard, and the wire from the ground (GND) terminal should connect to the bottom row of your breadboard.

4. Turn on your power supply.

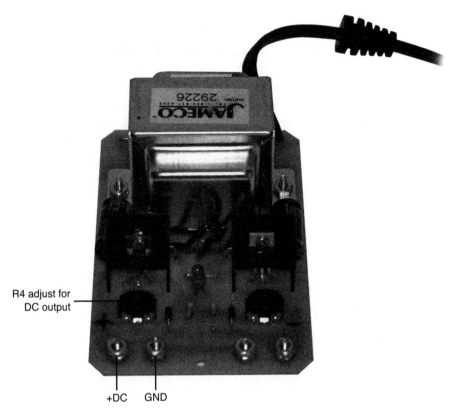

R4 adjust for
DC output

+DC    GND

*The +DC and GND terminals on your power supply.*

5. Using the following binary chart below as a reference, turn the DIP switches on and off to change the numbers on the display from 1 to 9. You are loading this data into the decoder by hand.

## Switch Inputs for Binary Values and Their Digital Output

| Switch 1 | Switch 2 | Switch 3 | Switch 4 | **BINARY** | Display |
|----------|----------|----------|----------|------------|---------|
| OFF | OFF | OFF | OFF | 0000 | 0 |
| OFF | OFF | OFF | ON | 0001 | 1 |
| OFF | OFF | ON | OFF | 0010 | 2 |
| OFF | OFF | ON | ON | 0011 | 3 |
| OFF | ON | OFF | OFF | 0100 | 4 |
| OFF | ON | OFF | ON | 0101 | 5 |
| OFF | ON | ON | OFF | 0110 | 6 |
| OFF | ON | ON | ON | 0111 | 7 |
| ON | OFF | OFF | OFF | 1000 | 8 |
| ON | OFF | OFF | ON | 1001 | 9 |

# Lab 15.2: Building a Decoder Circuit, Part 2

In the previous lab you used the DIP switches to send signals to the decoder. In this lab, you will remove the DIP switch and replace it with a timer and a decade counter. The timer will send out pulses, and on each pulse the decade counter will count up from 0 to $F_{hex}$ (hexadecimal) or 0 to $15_{10}$ (decimal) and then restart at 0. The frequency of the timer is set by the capacitors and resistors connected to the timer.

**Materials:**

Completed Part 1 lab

1 555 timer IC

1 7490 decade counter IC

1 100 KΩ resistor

1 1 MΩ resistor

1 10 µF electrolytic capacitor

1 0.01 µ capacitor

Jumper wire

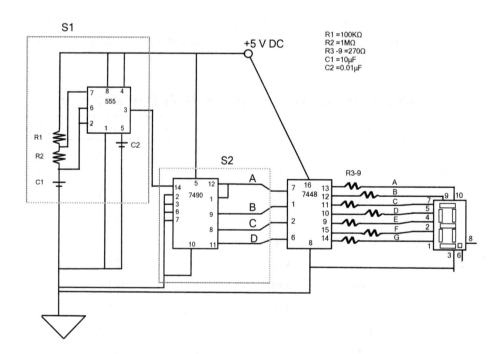

**Instructions:**

1. Before assembling your circuit make sure your power supply is off.

2. Remove the DIP switch.

3. Follow the schematic to add the timer IC (labeled S1) and the decade counter IC (labeled S2) as shown in the schematic.

4. Turn on your power supply. The numbers on the display should change every four seconds. When it is running, the counter goes from $0000_2$ to $1111_2$ (decimal numbers 0 to 15). With just a single display, numbers containing more than two digits (9 and above) are represented by symbols.

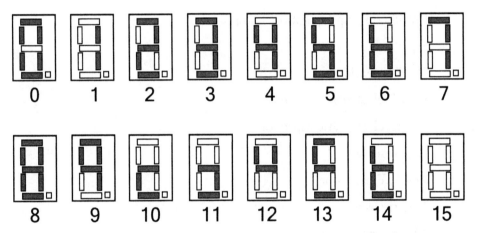

*Your display will show the digits 0-9, and then a series of symbols to represent the digits 10-15.*

## Lab 15.3: Guitar Amplifier

In this lab you will build a small, basic guitar amplifier. This project is based on a design by Runoffgrove.com and used with their permission.

As the components in this project are fairly inexpensive, you might want to have extras on hand in case you damage any parts while soldering them. But if you work slowly and deliberately, you should do just fine.

**Materials:**

1/4-inch mono audio jack

0.01 µF capacitor

LM386 op amp (operational amplifier)

8-pin socket

5K Ω potentiometer

100 µF electrolytic capacitor

220 µF electrolytic capacitor

10 Ω resistor

0.047 µF capacitor

25 Ω rheostat

9 V battery

9 V battery connector with leads

8 Ω speaker

Perf board has pre-drilled holes covering the entire board. There are solder pads on every hole, but none of the solder pads are connected to one another. You solder your components in place and then connect them to each other by soldering pieces of wire to make the connections.

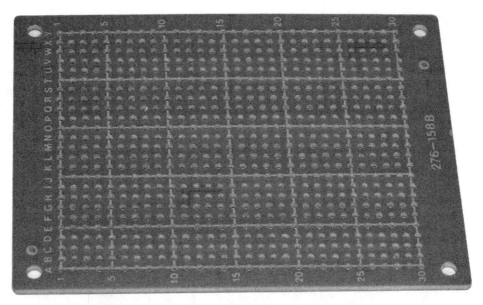

*Front side of perf board.*

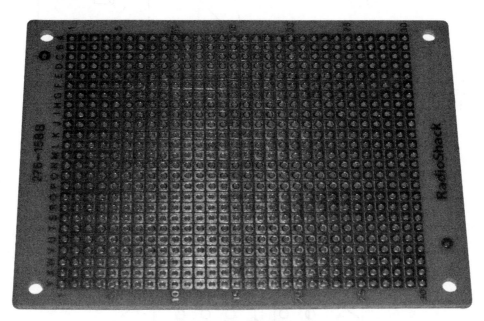

*Reverse side of perf board.*

**Instructions:**

1. Attach the 8-pin socket to the board. This is the most delicate part of the project.

2. Place the rest of the components on the circuit board as indicated by the schematic and diagram, but don't solder them yet. First, expose the last inch or so of wire on each of the leads that are marked as GND. Then twist the wires to make a connection between them all. Now you can solder them together.

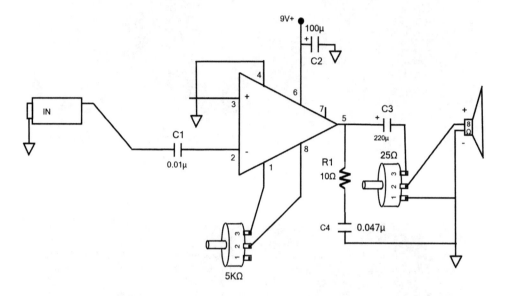

$\nabla$ = GND

*Schematic for guitar amp project.*

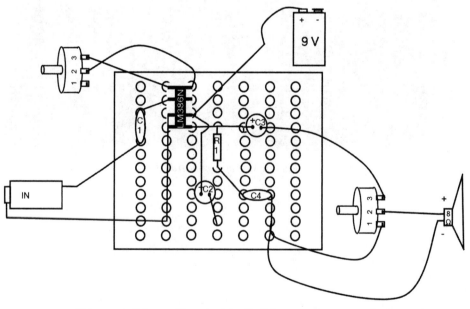

*Diagram of the perf board with all of the components attached.*

3. Use the other component leads to connect them to the appropriate pins of the socket, the speaker, and the other components.

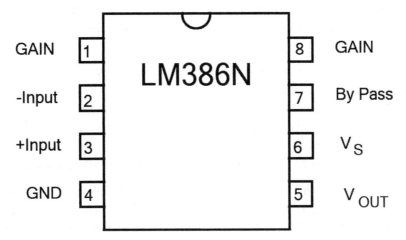

*Pin assignments of the LM386 op amp.*

4. Use some jumper wire to connect pins 3 and 4 to the interconnected ground leads. Connect the lead of the negative terminal of the battery connector to the ground leads. Connect the positive terminal lead to the junction of pin 6 and the 100 μ electrolytic capacitor.

5. Connect the LM386 to the socket and the 9 V battery.

*Completed guitar amp.*

6. To test your amp, pick up your guitar and connect your cord to the plug on the amp and give it a strum. Adjust the two potentiometers (5K $\Omega$ potentiometer is the gain adjustment and the 25 $\Omega$ rheostat is the volume adjustment) and listen to the effects. Be aware that nearby electrical equipment such as your soldering iron can cause interference. An enclosure will help to shield the amp from such unwanted noise.

As described the project does not have an enclosure or a power switch. There is a wide variety of enclosures on the market specifically designed for electronics projects. You might want to get creative and come up with your own enclosure, perhaps in a old cigar box or in a box constructed of plastic building blocks. Just be sure to use a nonconductive material (wood or plastic) with enough room inside and outside for all the parts and to be able to securely attach everything to protect against damage to your circuit and connections.

You can add a power switch by getting a SPST (single-pole, single-throw) switch. Cut the positive lead of the battery connector. Solder one end of the lead to one side of the switch and one end to the other side of the switch. You may want to add this after choosing an enclosure. This gives you the flexibility to have the switch in an aesthetically and functionally appropriate place.

# Memory

## In This Chapter

- Digital vs. analog storage
- Data and addressing
- Writing and reading
- Volatile and nonvolatile memory
- Storage media

If we had written this book 10 years ago, we would probably have titled this chapter "Computer Memory," but memory is now an essential element of most modern digital electronics. From the computers that run your car's various systems, to the digital picture frame on your desk, to the greeting card you pick up at the drug store, digital memory storage has made its way into all sorts of consumer and industrial products. Advances in chip design and manufacturing are leading to cheaper, bigger, more stable memory, requiring less power consumption every year.

While your beginner electronics projects may not involve a great deal of memory, if you do any further study of electronics or plan to create more complicated projects, you will need to use memory. Almost all modern electronics have at least a small amount of memory, and the dividing lines between computer and electronic devices are becoming more meaningless.

# Digital vs. Analog Memory Storage

Analog memory storage holds a physical representation of an analog signal; in other words, it holds the full continuous wave as opposed to the sampled stepped representation of a digital signal. Analog recording technologies have included magnetic recordings on tape and on wires as well as physical representations on disks of vinyl and wax. Analog storage is susceptible to degradation or damage either by magnetic interference or by physical damage to the storage medium.

Digital memory storage stores a sampled portion of an analog signal or digital signals. Digital signals stored on any medium are less likely to be degraded because of the nature of digital signals. Instead of analog's continuous wave, digital signals are composed of a series of either high or low voltages. Slight changes to a magnetic or physical storage medium are likely to fall within the tolerances of either a high or low voltage and so are much less likely to affect the digital signals.

## Parity Bits and Other Error Detection

Digital signals are still subject to error in transmission, data storage, and retrieval. Spikes or dips in voltage, electromagnetic interference, and other factors such as faulty hardware might introduce errors. This is not just an issue with digital signals; analog signals also can have errors introduced when transmitted, stored, or retrieved. Digital signals are, however, easier to "error correct." Error correction is a system that helps to identify errors so these errors can be filtered out. Error detection is simply the process of identifying errors; it doesn't involve correcting them.

Error detection was especially important in early computing when banks and other large businesses needed to trust that their information was accurate and error-free. For industry leaders to trust in the soundness of their digitally stored information, error correction methods needed to be developed.

Some of these methods just check for or detect errors (error detection) and some both detect and correct (error correction). Often these terms are used interchangeably, but it is good to know that these are actually separate actions.

One of the most basic error check methods is to add a *parity* bit. A parity bit indicates whether a number is even or odd. ASCII characters occupy seven bits out of eight to allow for the addition of a parity bit. Before a byte (which is 8 bits) is transmitted, the digits representing the data are added together and determined to be either even or odd. The parity bit is then occupied with either a 1 or a 0. When the byte is transmitted, the parity bit travels along with the data. When the data is received, the parity bit is checked and bytes that don't match the parity are discarded and an error message is created.

**Parity** is a relationship of oddness or evenness between numbers.

In telecommunications, a method for detecting and correcting errors is forward error correction (FEC), in which the data is transmitted more than once with an embedded code that satisfies a pre-established code.

Another error detection method used for lines of numbers that represent bank accounts or identifiers such as Universal Product Codes (UPC) or International Standard Book Number (ISBN) codes, a check digit can be used to protect against errors in human data entry or in transmittal. So-called "checksum *algorithms*" use a block of data that is compared at set events. Some checksum schemes allow for data correction in addition to error detection.

**DEFINITION**

An **algorithm** is a problem solving procedure or mathematical problem; usually it's an equation describing a relationship between the numbers in a string of numbers.

## Hexadecimal

Everything we have talked about so far when discussing digital circuits and memory has been linked to binary numbers. However, many computer designers and programmers use a base-16 numbering system called hexadecimal. Hexadecimal allows us to store more data in fewer bits.

Hex digits range from 0 to 9, and the additional digits are represented by the letters A to F. Each nibble (four bits) can hold a hexadecimal value that can represent up to 65,535 in decimal ($65535_{10}$, base-10). That same number would be $1111111111111111_2$ in binary (base-2) and $FFFF_{16}$ in hexadecimal (base-16). You can see that the hexadecimal (or hex) version would take up a lot less space in memory.

**WATTAGE TO THE WISE**

We are so used to the base-10 numbering system (digits 0–9) that we may not realize there are an infinite number of possible numbering systems (although only a few are in practical use). Binary numbers (digits 0–1) are base-2 systems and hexadecimal (digits 0–9 and A–F) are base-16 systems.

Subscripts can be used to indicate what base numbering system is being used. For example, $1101_2$, $1101_{bin}$, or $1101_{binary}$ all represent a binary number; $13_{10}$, $13_{dec}$, $13_{decimal}$ represent the decimal; and $D_{16}$, $D_h$, or $D_{hex}$ represent hexadecimal.

| Decimal - HEX - BINARY Conversion Code Chart | | | | | | | | | | | | | | | |
|---|---|---|---|---|---|---|---|---|---|---|---|---|---|---|---|
| Decimal | 0 | 1 | 2 | 3 | 4 | 5 | 6 | 7 | 8 | 9 | 10 | 11 | 12 | 13 | 14 | 15 |
| HEX | 0 | 1 | 2 | 3 | 4 | 5 | 6 | 7 | 8 | 9 | A | B | C | D | E | F |
| BINARY | 0000 | 0001 | 0010 | 0011 | 0100 | 0101 | 0110 | 0111 | 1000 | 1001 | 1010 | 1011 | 1100 | 1101 | 1110 | 1111 |

*Conversion chart for binary, digital, and hexadecimal numbers.*

# Data and Address

When using memory, you need to consider two different things: the actual information or data (*datum* is the singular version of *data*) and where that data are stored. The storage location is referred to as the data's *address*. The address includes both the address and the instructions for retrieving the information. In a computer, addresses are stored in the memory register, which is part of the central processing unit (CPU).

**WATTAGE TO THE WISE**

Just because you store something doesn't mean you can find it. Think of the dreaded junk drawer you may have in your kitchen or a basket of laundry. Each of those storage places can store things but do not have a system to find anything. It is not an efficient system.

Now think of a filing cabinet with files arranged alphabetically or a datebook that stores appointments chronologically. Each of those has an addressing system. If you know the system you can easily find what you are looking for. Memory addressing methods are like a filing system. It describes where data is stored so that it can be retrieved later.

There are two primary methods used in memory addressing: stack data architecture and dynamic memory allocation. With stack memory, data are stored in stacks of memory on a last-in, first-out basis. Data just keeps on getting stored without any check to see if enough space is available. It is fast and simple, but it is subject to error when more memory is allocated than is available; this is called a *stack overflow error*.

With dynamic memory allocation, memory is allocated as the programs run. Available memory is taken from the heap (a supply of available memory storage locations). This approach is subject to fragmentation of data, which happens when the available memory is scattered and only available in separate locations. This can slow the retrieval of the data when it is fragmented widely.

# The Von Neumann Computer Model

In the 1940s, when the very early modern computers called the Electronic Numerical Integrator and Computer (ENIAC) and Electronic Discrete Variable Automatic Computer (EDVAC) were being developed, computer scientists needed to design a system architecture for computers that could be used going forward so that the basic structure of a computer didn't need to be reinvented again and again. The model that became the standard going forward is known as the Von Neumann model.

**TITANS OF ELECTRONICS**

John Von Neumann (1903–1957) was a Hungarian-born mathematician who came to Princeton University in 1930 and was part of the founding faculty of the Institute for Advanced Study (along with other notables including Albert Einstein and Kurt Gödel).

The Von Neumann model owes much to the theories of Alan Turing and others, but he definitely had a major role in the adoption of the architecture as a standard. Von Neumann contributed to many fields from math to economics to mechanics to quantum theory, and many scientific and mathematical concepts bear his name.

The Von Neumann model (named after mathematician John Von Neumann) consisted of three main portions: the CPU, the input and output (I/O) subsystems (all of the I/O devices), and the memory. These building blocks are all connected by a system *bus*, the data bus, and the control bus.

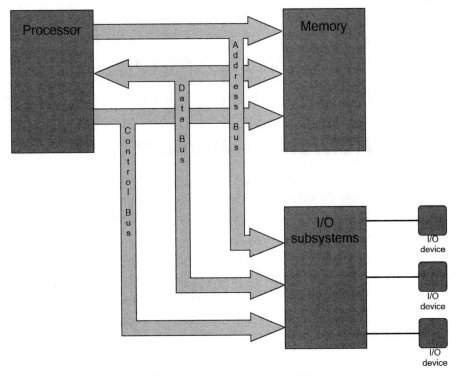

*Von Neumann computer model.*

**DEFINITION**

A **bus** in electronics is the electrical connection between components. In computing, the term refers to the transfer system of information between components.

## Memory Registers

Remember that data are stored in an IC called a memory register, which is part of the CPU. The two primary types of memory registers are the Memory Address Register (MAR) and the Memory Data Register (MDR). The MAR holds the location and the instructions; the MDR stores the actual data. When data are fetched, they are moved to the Memory Buffer Register (MBR) which is a holding position, then on to the Instruction Register (IR) where the specified programming instruction is performed.

# Writing and Reading

Storing data is called *writing*; retrieving data is called *reading*. Some memory doesn't allow for writing, and this is called read-only memory (ROM). For example, a CD-ROM is a stand-alone read-only portable media device.

Different storage media are then described as whether or not they are writeable. Some CDs and DVDs are labeled RW. That means that they are both readable and writeable. Other media is read-only.

The firmware (the embedded instructions that coordinate the operation of the computer's internal processes) of a computer is typically ROM. It contains the basic operating instructions and file structure for the computer. Some computers use EPROM to store their firmware. EPROM is erasable, programmable ROM. It is still considered ROM, however because it is infrequently rewritten and is not available for general storage on the computer.

# Volatile and Nonvolatile Memory

Volatile memory refers to memory that is power-dependent. If power is not supplied to the memory, it loses its data. This data loss is not necessarily instant, as the power is held in the capacitors, but volatile memory is eventually lost. Nonvolatile memory is able to maintain data even when there is no power.

An example of volatile memory is random access memory (RAM), which is the working memory of your computer. It is called random access because it doesn't need to be accessed sequentially. In modern computers, however, ROM is random access as well. The important distinction is that, generally, RAM is volatile memory whereas ROM is not.

# Storage Media

Several types of media are available for storing memory. These include magnetic storage on tapes; magnetic storage on disks (hard drives); optical storage on removable disks; and flash (which is a type of electrical erasable programmable ROM, or EEPROM), and other solid-state memory (that is, media storage that doesn't require any moving parts).

Each storage type has drawbacks. Magnetic storage can be corrupted by exposure to magnetic fields (either electromagnetic or natural magnets). Hard drives can suffer from mechanical failure. Optical storage can degrade because of the chemical characteristics of the storage disks themselves. Flash memory can degrade after multiple reprogramming and is expensive relative to other memory types. Each type can be protected against these threats, and flash memory is becoming more affordable per byte of memory. New types of solid-state memory are being developed, each trying to become a near-perfect *universal memory*.

> **DEFINITION**
>
> **Universal memory** is a goal toward which computer scientists are working. The ideal memory would be both affordable to produce and energy-efficient. It must be fast, nonvolatile, and resistant to magnetic interference.

## The Least You Need to Know

- When an analog signal is stored, the entire signal wave is stored. Digital signals, which are stored by recording low and high voltages, are less likely to degrade.
- Two types of memory are stored in computers: the data themselves and the address where the data are stored.
- Storing memory is called writing; retrieving memory is called reading.
- Volatile memory is memory that isn't preserved when not powered; nonvolatile memory is stored even if no power is present.
- There has been a continual evolution in storage media moving toward the ideal of universal memory, which is memory that is affordable to produce, nonvolatile, fast, and resistant to magnetic interference.

# Microcontrollers

## In This Chapter

- Microcontrollers vs. computers
- Understanding how microcontrollers are made
- Programming microcontrollers
- Shopping for microcontrollers

Microcontrollers (MCUs or µCs) are self-contained, embedded computers. When a computer or device is embedded, it means that it is integrated into the operation of that device and isn't available for purposes outside of that device. Everyday devices from microwaves to your television remote have microcontrollers that enable these devices to execute commands based on inputs and generate outputs that are customized to the operation of that particular device.

## What Are Microcontrollers?

Microcontrollers are computers, but they do not have a lot of memory because they are meant to perform a discrete set of tasks. Instead of following a Von Neumann architecture (see Chapter 16), they are designed using the Harvard architecture, which has separate buses for instructions and data storage so that operations can be completed more quickly.

Notice that Harvard architecture has two separate memory locations—one for instructions and another for data. Programmers can use the two separate memory locations to their advantage by using different address schemes for the data memory and the instruction memory. One could use a 16-bit scheme and the other could use an 8-bit scheme. In regular computers, the address scheme would have to be consistent for both types of memory (instructions and data). Having two address schemes means that the microcontroller designers can use faster machine level code in one memory location of the microcontroller for the microcontroller's firmware and more user-friendly programming tools when giving instructions in the other memory location.

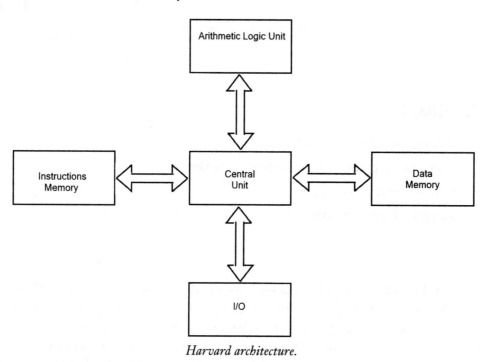

*Harvard architecture.*

**HIGH VOLTAGE!**

Don't confuse microcontrollers with microprocessors. Microprocessors are integrated circuits (ICs) that serve as the central processing unit (CPU) for a computer.

Microcontrollers are complete computers with their own memory, input/output (I/O) ports, and a CPU. In practice, microcontrollers are part of an embedded computer system with various I/Os such as sensors, switches, and displays. Because they are designed to execute specific tasks and the programs tend to be much smaller, they do not require a lot of processing power or memory.

Just because microcontrollers seem simpler than computers doesn't mean that they are just toys. Our project in upcoming chapters is a basic robot with a few sensors, but even a single sensor allows for many useful and powerful applications. Once you understand how to work with microcontrollers, you can go on to construct more sophisticated projects, such as automating an aspect of your home (security, light timing, feeding of your pets), creating a holiday light display, or using a global positioning system (GPS) to monitor the flight and altitude of your rocket or remote-control (RC) airplane. The types of projects that a microcontroller can execute are limited only by your imagination.

# Microcontroller Components

Typical microcontroller components include a CPU, serial I/O ports, memory (both volatile for data storage and read-only memory [ROM] or flash memory for instructions), and usually an analog-to-digital converter (ADC). Other components can include timers, clock generators, and event counters.

## Inputs/Outputs

When choosing a microcontroller, be sure that there are an adequate number of I/Os for your project and that they are of suitable types. Most should have both analog and digital I/Os. A computer-compatible I/O for programming is useful as well. Most use a universal serial bus (USB) or micro USB connector. If you have a particular project in mind you can research how many I/Os that project will require.

Ajay V. Bhatt is the creator of the USB I/O. You may have seen him on Intel's "rockstar" commercials, or at least an actor portraying him. Coming from a middle-class family in India, he moved to the United States to study at the City University of New York. As a consumer himself, he became frustrated at having different connectors for his computer and saw the need for a universal connection device. He and his team at Intel developed the USB as a universal connector for computers and other devices and peripherals, as well as specialized software to make connections. Now USB connectors are used on all sorts of devices.

He also was part of the team that developed other I/O interfaces including the Accelerated Graphics Port (after he saw a need) and the PCI Express, two interfaces for graphics cards. Mr. Bhatt has appeared on *The Tonight Show with Conan O'Brien* and in *India GQ*, where he was listed as one of "50 Most Influential Global Indians."

# Programming Microcontrollers

Microcontrollers must be coded with instructions that the CPU understands. The most basic instructions must be in machine *code* or machine language, which is made up entirely of zeros and ones. As machine language is very unwieldy, assembly language is used to translate basic commands into machine language. Assembly language is considered a low-level programming language. Low-level languages are more widely applicable to any computing environment, but they are much more difficult to program. Programmers usually use high-level programming languages and then use a compiler to translate the *source code* of the high-level program language into assembly language.

**DEFINITION**

When programmers talk about **code,** they are referring to instructions in a program, not just machine code. Coding means programming. **Source code** is the instructions in a particular language; generally, it is not written in a user-level language.

Most microcontrollers can be programmed using user-friendly, high-level computer languages such as the C programming language family (C, C+, C++, and C#) or BASIC (Beginner's All-Purpose Symbolic Instruction Code) and its various *dialects* (especially Visual Basic).

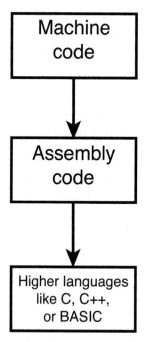

*The hierarchy of programming languages.*

**Dialects** in programming languages are just like dialects in any language—they are slight differences in the language from region to region and group to group. A dialect of a programming language reflects its programming environment and its purpose.

## Microcontrollers for Hobbyists

Many different models of microcontrollers are designed for hobbyist uses. When you choose one, you want to make sure that it meets your needs. Things to consider are the number and type of I/O ports, the programmability (the language used and whether or not its software is easily reprogrammed), the amount of memory available for instructions, and the cost.

Some popular microcontrollers include:

- Microchip PIC microcontroller: one of the earliest affordable microcontrollers; it uses a freeware development package called MPLAB and the C programming language.

- Atmel AVR microcontroller: has flash memory for program storage. AVRs have been used in many automotive applications and even with Microsoft Xbox controllers.

- Intel 8051 and 8052: these chips are used in many engineering schools as introductory microcontrollers.

- BASIC Stamp: very popular with hobbyists because of its use of the simple BASIC programming language.

- Open source hardware microcontrollers (Arduino, Netduino): these so-called open source hardware microcontrollers have seen a rapid rise in popularity with hobbyists due to the balance between ease of use and adaptability to a wide range of products.

## Open Source Hardware

Following on the ideas of open source software, several microcontrollers are part of the open source hardware community. Open source software (or free and open source software) is software for which the source code is available so others can review and modify it for their own use. Open source hardware is hardware that publishes its schematics, printed circuit board (PCB) design, embedded programming code, and the like.

Two microcontrollers have been developed under the open source hardware scheme (and the software scheme for programming each): the Arduino and Netduino electronics platforms.

## The Arduino Microcontroller Platform

The Arduino platform was developed by two Italians, Massimo Banzi and David Cuartielles, as an inexpensive and easy-to-use microcontroller (and control software) for students and hobbyists to create and test prototypes for embedded devices of their own design or to create projects from others' designs. The creators named the Arduino after a favorite bar in Ivrea, Italy.

The programming language is based on C#, and the processor is an Atmel AVR microprocessor. It includes a bootloader, which allows the user to upload to the flash memory easily. The bootloader is designed to work with different *shields* (add-on modules or *daughterboards*) that can easily be connected directly to the CPUs.

**DEFINITION**

A **daughterboard** is a circuit board that connects directly to the CPU or motherboard. It doesn't use wires or computer buses to connect, but instead integrates with the main circuit board.

The Arduino is a cross-platform system, meaning that it can work in Windows, Linux, or Mac operating system (OS) environments.

Many versions of the Arduino microcontroller are on the market. Because it is open source hardware, anyone can use their designs to create their own (with some restrictions). There are many online user forums for Arduino-based projects.

## The Netduino Microcontroller

Our project is based on the Netduino microcontroller. It, too, was developed under the open hardware scheme. Netduino is designed for beginners and hobbyists but also for commercial project prototyping. It is based on the .NET Micro Framework, a version of Microsoft's .NET, which is an operating environment for low-memory embedded devices. (See Appendix F for information on running the .NET platform with a Mac operating system.)

## The Least You Need to Know

- Microcontrollers are self-contained, embedded computers. They rely on a different architecture than standard computers and have smaller memory available.
- Microcontroller components include a CPU, memory (volatile and nonvolatile), inputs/outputs, and usually an analog-to-digital converter.
- Programming microcontrollers involves low-level computer instructions or machine language.
- Several microcontrollers have been designed for use by hobbyists and students. The project in this book works with the Netduino microcontroller.

# Constructing a Robot

To get started building your very own robot, you first need to do a little shopping. Once you have all your supplies, you'll need to do some basic programming using your home computer and a microcontroller, attach some wheels, and give it some motor power and control.

Then we look at all of the sensors that we can use to give your robot its many jobs. At their most basic level, sensors sense things. Imagine anything that can be sensed—light, temperature, radio signals, gases, and a whole lot more—and there is probably a sensor that detects it. We'll learn the basics of transmission, the steps, and the methods. The information that the sensors provide can be fed back to your robot and, with your help, use the information to make decisions. You will use a very sophisticated ultrasonic sensor to detect objects and make your robot do something very unsophisticated: dance.

# Building Your Robot

## In This Chapter

- Shopping for supplies
- Getting the software
- Writing the code
- Downloading programs to your microcontroller

Time to get your geek on: first by shopping, then by building a robot—a machine that follows a set of instructions to perform different tasks without human assistance. By the time you're done with this chapter, you'll have your robot up and running and even performing a trick or two. You'll also have the knowledge you need to try your own programs and add some capabilities on your own.

## Shopping for Your Robot

All of the parts and components you'll need for projects in the rest of this book are available from a wide variety of sites online. We prefer to order from SparkFun Electronics (www.sparkfun.com) because it provides helpful advice about using its products and offers online tutorials. See Appendix F for more online shopping recommendations.

Your shopping list should include the following:

Netduino Microcontroller (SKU: DEV-10107)

Ardumoto Motor Driver Shield (SKU: DEV-09815)

2 Micro Metal Gearmotors 100:1 (SKU: ROB-08910)

Wheel 42 × 19mm X2 (SKU: ROB-08899)

Micro Metal Gearmotor Bracket Extended (SKU: ROB-08896)

MaxBotix Ultrasonic Rangefinder LV-EZ1 (SKU: SEN-00639)

Wall Adapter Power Supply—9 VDC, 650 mA (SKU: TOL-00298)

9 V to Barrel Jack Adapter (SKU: PRT-09518)

Break Away Headers—Straight (SKU: PRT-00116)

Breadboard Mini Self-Adhesive (SKU: PRT-07916)

Break Away Headers—Right Angle (SKU: PRT-00553)

**Option A Base:**

Sturdy cardboard box top at least 6" × 9".

Toy wheels with axles (approximately 1½" in diameter, repurposed from any used toy)

**Option B Base:**

Universal Plate Set (SKU: ROB-10016)

Toy Tires—Basic (SKU: ROB-00423; note: this is a set of four but you'll only use two for the front tires)

Tamiya RC Rein. Freewheel Axle Set—M-Chassis (SKU: #54183; available from RC hobby shops or online from the manufacturer)

You should be able to purchase all of these items for under $175.

# Get the Software You Need

The software you need to program your microcontroller—Microsoft Visual C# 2010 Express—is available for free download. Microsoft Visual C# 2010 Express is part of the .NET Framework, which is an integrated development environment (IDE) that provides support for programming in multiple languages, debugging, and much more. The .NET Framework is a set of standards that enables programmers to develop programs in one of several programming languages and to have their programs run on any processor that is designed to work under these standards. Working outside of this framework requires knowing the particular demands of each processor, often with its own quirks and "dialect" of even the standard languages.

We will be working in the C# (pronounced C sharp) language for .NET Framework. This will require layering support for the .NET Micro Framework onto the .NET which expands the environment from computer processors to work with microcontrollers.

The Netduino microcontroller is specifically designed to take advantage of the .NET Micro Framework. These tools are designed to work with Windows operating systems. On each linked page, you will find any minimum requirements. (See Appendix F for solutions if you have a Mac operating system.)

Download and install the following software on your computer in this order:

1. Microsoft Visual C# 2010 Express:
   www.microsoft.com/express/downloads/#2010-Visual-CS

2. .NET Micro Framework 4.1 software development kit (SDK):
   www.netduino.com/downloads/MicroFrameworkSDK.msi

3. Netduino SDK v4.1.0 32-bit or 64-bit, depending on your computer's operating system:
   www.netduino.com/downloads/netduinosdk_32bit.exe
   or
   www.netduino.com/downloads/netduinosdk_64bit.exe

# Connecting Your Netduino and Getting to Work

The Netduino microcontroller has 14 digital input/output (I/O) pins, 6 analog input pins, and power ground (GND) pins; they are all labeled on the board. There is also a programmable button and a light-emitting diode (LED). You will be working with the LED in this section.

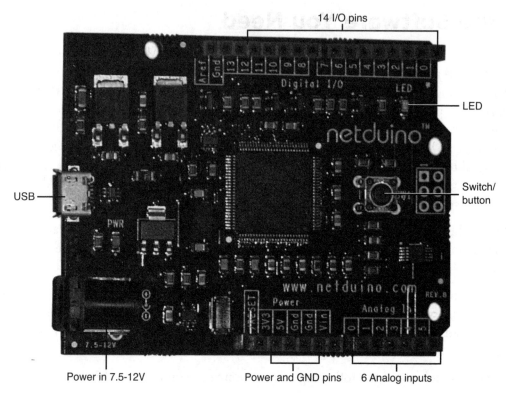

14 I/O pins

LED

Switch/
button

USB

Power in 7.5-12V     Power and GND pins     6 Analog inputs

*The various parts on your Netduino microcontroller.*

Connect the Netduino microcontroller to your PC with the Micro universal serial bus (USB) cable that it comes with. Your system should detect it and set up all the drivers.

Now you are ready to start your first program. This program makes an LED on the system board blink. In the process of creating this program, you will be introduced to working with the C# programming environment and some of the basic programming we will expand on later in the book.

**WATTAGE TO THE WISE**

The program running on your computer is called a compiler. This will take the code you write in C# and translate it into a machine-friendly language. You will only see the C# language code.

1. Start Microsoft Visual C# 2010 Express.

2. Choose **New Project.**

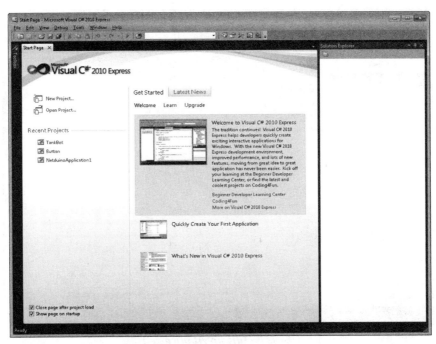

*The start page for Microsoft Visual C# 2010 Express. Choose **New Project.***

3. Click the triangle next to Visual C# in the upper-left corner to expand the list. Choose **Micro Framework.** In the project type list, choose **Netduino Application** and name the project **LED Blink.**

*Click the triangle next to Visual C# in the upper-left corner to expand the list.*

*Choose **Netduino Application**.*

4. Click **OK.** You should now see the following screen.

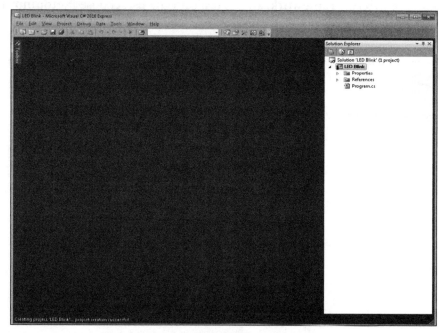

*You should see the LED Blink project on the list.*

5. Double-click **Program.cs** in the list on the right. This creates the proper form to start writing your code.

6. You can now start typing code immediately after "//write your code here":

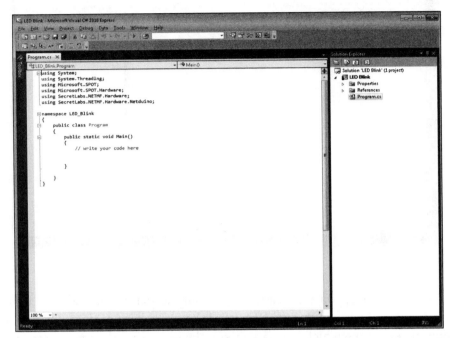

*Start typing code here.*

```
OutputPort led = new OutputPort(Pins.ONBOARD_LED, false);
  while (true)
  {
      led.Write(true);
      Thread.Sleep(250);
      led.Write(false);
      Thread.Sleep(250);
  }
```

*Your code as it should be entered.*

In the first line of code, you are creating a new output, naming it "led" and telling the program to use the pin named ONBOARD_LED. You may have noticed when typing that this list of pins appears from which you can choose the pin rather than typing in the complete name of the pin.

Adding "false" after the pin name sets the LED to "off" as its default state.

The "while (true)" statement tells the LED what to do when it is on.

The "led.Write(true);" statement turns the LED on.

The "Thread.Sleep(250);" statement keeps the LED on for 250 milliseconds (ms).

The "led.Write(false)" statement turns the LED off and the line "Thread.Sleep(250)" sets the time interval at 250 ms.

**WATTAGE TO THE WISE**

When writing code, you will get prompts on what the compiler thinks you may need next; this is called *intelli-sense*. A short list of possible commands to use next will pop up. You can select this command with the arrow keys and then press the **Tab** key to insert it into the code you are writing.

# Downloading to the Netduino Microcontroller

Now let's download the code to your Netduino microcontroller:

1. From the menu bar, choose **Project,** then **LED Blink Properties.**

2. Under **Target framework,** select **.NET Micro Framework 4.1,** and under **Transport,** select **USB.**

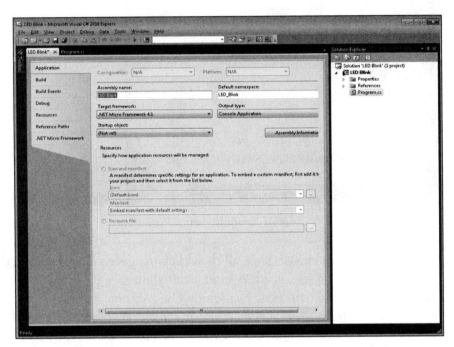

*Under **Target framework,** select .NET Micro Framework 4.1.*

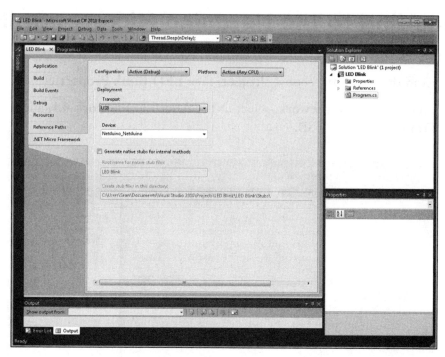

*Under **Transport**, select **USB**.*

3. Under the file menu choose **Save all** then click the green play arrow on the top menu bar or press **F5** to *debug* the program. If your code is correct it will load the program you entered onto your Netduino. The LED on the microcontroller will be blinking. If you have any errors or *bugs*, the screen on your computer will highlight any bugs that need to be corrected. Check that your code is typed correctly, as typos are the most common errors.

**DEFINITION**

A **bug** is an error in programming that causes unintended consequences. To **debug** a program is to find errors and correct them.

Blinking LED

*If you installed the program correctly, your LED should be blinking.*

## The Least You Need to Know

- To build a microcontroller-controlled robot you need a microcontroller, a motor, and wheels. Other accessories can include sensors and precision control components.
- The .NET framework is a set of standards that work with several programming languages and are designed to work with any processors that support those standards; this is in contrast to processor-specific programming requirements.
- Downloading your program via the Micro USB from your computer to the Netduino is controlled from the Visual C# compiler on your computer.
- When programming with Visual C# 2010 Express, there is a built-in debugger to catch mistakes in your code.

# Motors and Controllers

## In This Chapter

- Understanding brushed and brushless direct current (DC) motors
- Using stepper and servo motors
- Controlling motors using pulse width modulation (PWM)
- Controlling motors using H-bridge circuits

To get your electronics to do some work, you need to enlist the aid of motors. You can use electronic controls to regulate the current to drive the motor. Choosing the right motors and control systems enables you to move, make, and drive your designs in a deliberate and effective manner.

It is easy to get lost in all of the discussions of new digital devices and think that electronics is all about computing and home entertainment, but electronics is so much more than these gadgets and consumer products. Electronics play a large role in manufacturing, automotives, energy (production, delivery, and heating/cooling), medicine and yes, computers and home entertainment. If you are planning to pursue a career in electronics, you need to understand how electronics interact with things that move. If you want to build remote control cars, airplanes, or bomb disposal robots, you need to make them move—and motors are what make things move.

## Brushed DC Motors

A basic brushed DC motor operates through the use of a commutator, brush, and armature assembly. A magnet, called a *stator*, is fixed in place, and the armature is arranged so that it can revolve inside the stator's magnetic field. In Chapter 12 you learned about induction, and how a conductor inside of an electromagnetic field can generate an AC electrical current. In a motor, this process takes place when the armature (which is a conductor with a wire attached) revolves inside the magnetic field of the stator.

We now have a wire that is conducting current. When a wire is carrying a charge, it creates its own electromagnetic field. When the armature (think of it as the motor's arm) has a charge applied by a DC power source, it rotates 180° inside the larger magnet because of the interaction of the magnetic fields (like charges repel and opposite charges attract, causing an 180° spin). This rotational force is called *torque*.

**DEFINITION**

**Torque** is a rotational force around an axis or pivot point. It is sometimes referred to in mechanical engineering as *moment* or *moment of force*.

The *commutator* is a ring around the armature's axis. It has two gaps. DC power is applied to the commutator as it rotates through connections made by brushes (one on each end), which are like small brooms made of conductive material. The gaps on the commutator interrupt the power applied to the armature because the brushes don't make contact with the commutator.

The interaction of the magnetic fields creates a continuously reversing current (alternating current, or AC) because of the 180° rotation, and creates a continuous torque. The gaps in the commutator interrupt the electrical current at the point where the polarity would shift, so it continuously switches the direction of the current.

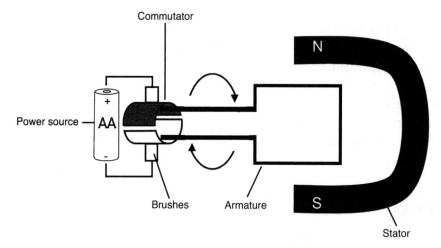

*Diagram of a brushed motor.*

The preceding diagram depicts the elements of a brushed motor, but in practical applications, there are more than two points of contact between the brushes and the commutator. The reversing polarity and the interaction of magnetic fields yields torque on the armature. This rotation powers the gear of the motor.

# Brushless DC Motors

Brushless DC (BLDC) motors operate without a commutator or brushes. Instead, they use electronic controls to apply reversals of polarity to the armature.

The lack of a mechanical commutator and brush system reduces the motor's weight, which translates into lower power consumption and less friction. In other words, these motors are more efficient. The downside is that the electronic controls make BLDCs more expensive than conventional brushed motors.

The current-to-torque ratio in brushless motors is linear. This means that proportional increases in current will result in proportional increases in torque. There isn't a loss due to friction, which affects the performance of brushed motors. This makes BLDC motors the preferred choice when precision applications are involved. There is also less opportunity for sparking as in brushed motors and there is a significant reduction in the mechanical noise.

BLDC motors are used in many consumer electronics applications, including computer hard drives and CD/DVD drives.

## Stepper Motors

A stepper motor is a brushless DC motor that is controlled by the interaction of multiple electromagnetic points (or steps) turned on and off to move the rotor in a highly controlled fashion. Stepper motors are the motor of choice in precision environments, where control is the most important factor. Stepper motors drive things as diverse as hard drives to aerospace applications.

There are three principal types of stepper motors. The oldest is the variable-reluctance–type motor with a large, iron-toothed gear set inside a series of electromagnets. The magnets can be powered to create rotation.

A second version of the stepper motor is the permanent magnet type. Instead of using a toothed rotor, it has a large permanent magnet constructed to have alternating strips of north and south polarities. Electromagnetic fields are created by two external terminals, which cause rotation between the stepped strips of the magnets.

Hybrid stepper motors are a mix of variable-reluctance and permanent magnet motors. In these motors, a permanent magnet rotates between the terminals but the rotor also has teeth to allow for more precise control of the steps.

## Servo Motors

A servo motor is a brushless motor that provides feedback to the motor control. The typical servo motor has an actuator arm that broadcasts its angle of rotation back to the control system. The feedback allows for error correction and real-time instructions. Servo motors can be any of the various motor technologies just described—it is the ability to provide feedback to the system that defines them as servo motors. One popular use of servo motors by hobbyists is in robotics and radio-controlled airplanes.

Servos are usually controlled by three leads: the power, the ground, and a control lead. The control lead transmits a signal that relates to the angle of an actuator arm. That signal is fed back to the control unit. The length of the pulse indicates the angle that the actuator is rotated, usually with a cap of 180°.

For many hobbyists, servo motors are their go-to motor because of how they are controlled and their flexibility of use. Servo motors are widely adaptable and generally affordable. Fans of shows like *Mythbusters* or who have seen any robotics competition will see the important role that servo motors provide when it is time to combine design, power, and agility.

> **WATTAGE TO THE WISE**
>
> Mechatronics is the field of study at the intersection of motor technology and electronics control. It is a synthesis of the fields of mechanical engineering, electronics, computing, and control theory. Smart systems require that students and professionals understand the theory of each of these areas. Your study of electronics is preparing you for this next generation of technology.

# Controllers

Be it a simple switch or a dedicated microcontroller, all motors need a control. In DC motors, if you apply a voltage, the gear will spin; if you reverse the voltage, you reverse the direction of the spin. To control a motor beyond these basic movements requires some control system.

You can control a motor's speed by managing the amount of current provided. The torque of a motor is directly related to the current driving it, so controlling the voltage through varying resistance or switches can manage the amount of current provided.

## Pulse Width Modulation

The use of pulse width modulation (PWM) is an effective method of motor control. By turning electronic switches off and on, the user controls the amount of power (voltage × current) to a motor. Each on and off pulse is set close enough together that the load is not perceptibly turned off. The percentage of off portions of the wave (when the switch opens the circuit and provides no power) compared to fully on (when the switch closes the circuit and the full current is flowing) is called the duty cycle. If the pulses are equally on and equally off, the duty cycle is said to be at 50 percent. Fans, simple heaters, and light dimmers all operate with PWM.

You can use pulse width modulation for speed control. A 30 percent duty cycle would provide a moderate amount of power to the motor and would drive the motor to 30 percent of its capacity. A 100 percent duty cycle would be full throttle on the motor. Most motors benefit from lighter duty than full throttle because of the reduced likelihood of wear and tear; in addition, lower duty cycles use significantly less power.

## H-Bridge

An H-bridge circuit is often used to control DC motors by reversing polarity. It consists of four switches that are turned on and off in different combinations that each yield different results.

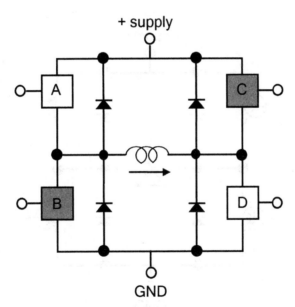

*An H-bridge symbol on a circuit diagram; A, B, C, and D are switches.*

Depending on what switches are opened or closed, you can control the motor's direction, apply a brake, or let the motor power off without braking.

## Settings for an H-Bridge Circuit

| A | B | C | D | Results |
|---|---|---|---|---------|
| 1 | 0 | 0 | 1 | Motor moves right |
| 0 | 1 | 1 | 0 | Motor moves left |
| 0 | 0 | 0 | 0 | Motor powers off without braking |
| 0 | 1 | 0 | 1 | Motor brakes |
| 1 | 0 | 1 | 0 | Motor brakes |

The other possible combinations of open and closed switches (such as 1111 and 1100) will result in short circuits.

Most H-bridge circuits are built with semiconductor transistors (either bipolar junction transistors [BJTs] or field effect transistors [FETs]) to protect against potentially damaging flyback voltages. A flyback voltage is like a switch bounce—the mechanical force of the motor turning on and off can generate a voltage. For this reason, diodes are usually included in the circuit.

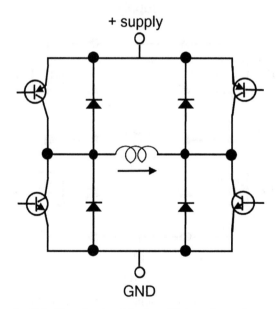

*An H-bridge circuit with the addition of transistors.*

## The Least You Need to Know

- Brushed DC motors convert current to torque through the interaction of the electromagnetic fields of an armature connected to a DC power source and a magnet called a stator. The brushes connect to a ring called a commutator, which has gaps to allow the current to alternate polarity.
- Brushless DC motors use methods of charge interruption other than a commutator and brush assembly, usually electronic controls.
- Stepper motors are brushless motors that operate by having intermittent connections between a conductive toothed gear and an electromagnet.
- Servo motors are motors that provide feedback to the controls.
- Motors can be controlled using pulse width modulation (PWM), with varying widths representing different degrees of rotation.
- Motors can be controlled using H-bridge circuits, which are a series of open or closed switches.

# Getting Your Robot Moving

## In This Chapter

- Mounting your motor and microcontroller
- Adding power
- Programming your robot to start and stop
- Controlling your robot's speed

You can't have a robot that just sits there. Let's get things moving by mounting a motor, adding a power supply, and giving the robot some controls. We'll start with a few commands, but the tools are in your hands if you want to add more.

# Get Your Motor on Board

Now it's time to add features to your robot to get it moving.

## Assembling the Motor Driver Shield

### Materials:

Soldering iron

Solder

Circuit board holder

The parts included with the Arduomoto Motor Driver Shield:

> Motor driver shield
>
> Shield board
>
> 2 6-pin stackable headers
>
> 2 8-pin stackable headers
>
> 2 2-pin screw terminals

**Instructions:**

Sparkfun has a great online tutorial on putting the motor driver shield together with color photos and soldering tips. You can find them on the product page on their website (www.sparkfun.com). Follow their instructions and assemble the shield.

After assembling the motor shield you will need to make one small change. This shield was designed for the Arduino microcontroller, and we are using a Netduino microcontroller; some of the output pins are different.

The Netduino uses I/O pins 6 and 9 for PWM (pulse width modulation) but the input for PWM on the motor shield is on 3 and 11. You will see them labeled on the board as PWMA and PWMB. This is an easy fix:

1. On the bottom of the board, trim off pin leads 3 and 11.

2. Next, use two small jumper wires on the top side. Use one to connect 3 to 6 and the other to connect 9 to 11.

After these changes, the output from the Netduino will be connected to the right input on the motor shield.

## Building a Platform and Mounting the Parts

For our robot, we used a cardboard box top as a base on which to mount all of the parts and repurposed a set of wheels from a toy truck we had lying around the house, described as Option A in the shopping list in Chapter 18. Option B uses the Universal Plate set, tires, and an axle. You can be inventive here. It is your robot, so feel free to be creative with your design. You can always take it apart and rebuild it.

# Power It Up

Initially, power for your Netduino microcontroller came from your computer via the Micro universal serial bus (USB) cable. This provided enough power for programming the device and making the light-emitting diode (LED) blink (see Chapter 18).

To give your assembly more power to drive the motor, you will need to connect the power supply to an alternating current (AC) outlet. Until your robot is ready to roll, prop up the entire assembly so that the wheels don't make contact with any surface. This way, you can make sure everything works properly before sending your robot out into the world.

*Robot with its wheels off the ground.*

Of course, your robot won't go too far connected via a cord to a wall outlet. When you are ready to let it go, you will need to provide 7.5–12 VDC. A 9 V battery will work just fine as well, but it will drain very quickly, so consider using a rechargeable battery.

*Barrel jack and battery connector cable.*

# Programming Your Robot to Start and Stop

Now let's write some C# code to control the motors. You will set and name the output ports and then have them go forward for 500 milliseconds then backward for 500 milliseconds and repeat until you disconnect the power.

1. Start Microsoft Visual C# 2010 on your computer. Choose **New Project** just as you did in Chapter 18.

2. Open the file **project.cs** and below the phrase "//write your code here," type the following code:

```
OutputPort Motor1 = newOutputPort(Pins.GPIO_PIN_D12, false);
OutputPort Motor2 = newOutputPort(Pins.GPIO_PIN_D13, false);
while (true)
{
  Motor1.Write(true);
  Motor2.Write(true);
  Thread.Sleep(500);
  Motor1.Write(false);
  Motor2.Write(false);
  Thread.Sleep(500);
}
```

3. Be sure to change the project properties to use the USB transport and the set device to Netduino, as shown in the following image.

**WATTAGE TO THE WISE**

You can disconnect the Micro USB cable after you have successfully uploaded the code to the Netduino microcontroller. It will hold the code until you reprogram it.

4. Press **F5** to upload the project to the microcontroller.

5. Reconnect the power cord, keeping the wheels elevated. Watch the motor go forward and in reverse. The code you entered tells the motor what to do.

6. Disconnect the power cord when you are sure that everything is functioning.

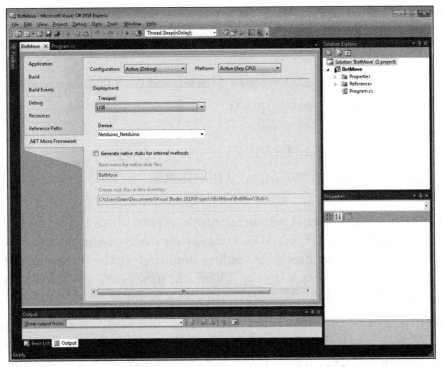

*Set your transport to USB and set the device to Netduino.*

**WATTAGE TO THE WISE**

After you load a program onto your Netduino microcontroller, the program will run as soon as you supply power to the microcontroller.

# Identifying the I/O Pins

Each input/output (I/O) pin on your Netduino microcontroller has a particular function, as follows:

> Digital pins 0 to 1: UART 1 RX, TX
>
> Digital pins 2 to 3: UART 2 RX, TX

I/O pins 0 through 3 are UART (universal asynchronous) RX (receiver), TX (transmitter); they send and receive I/O pins that are used for serial communications.

> Digital pins 5 to 6: PWM, PWM

I/O pins 5 and 6 are used for PWM (pulse width modulation).

> Digital pins 7 to 8: UART 2 RTS, CTS

I/O pins 7 and 8 are UART RTS (request-to-send) and CTS (clear-to-send).

> Digital pins 9 to 10: PWM, PWM

I/O pins 9 and 10 are also used for PWM.

> Digital pins 11 to 13: SPI, MOSI, MISO, SPCK

I/O pins 11 to 13 are SPI (serial peripheral interface) pins, which can be programmed as MOSI (master output/slave input), MISO (master input/slave output), and/or SPCK (serial clock). These pins are used mainly for sending commands to the Arduomoto Motor Driver Shield. When you set the output state to TRUE the motors will go forward, and when you set the output state to FALSE the motors will go backward.

All of these pins can do a lot more than what we describe here, but we will keep it simple for now. What you do need to know is that the I/O pins have specific functions. For example, you cannot program pin 11 to send PWM signals, but you can use pin 5, 6, 9, or 10 to send PWM signals.

## Adding Speed Control

You can now start to add the speed control program code:

```
PWM pwm1 = new PWM(Pins.GPIO_PIN_D6);
PWM pwm2 = new PWM(Pins.GPIO_PIN_D9);
OutputPort Motor1 = new OutputPort(Pins.GPIO_PIN_D12, false);
OutputPort Motor2 = new OutputPort(Pins.GPIO_PIN_D13, false);

while (true)
{
  pwm1.SetDutyCycle(30);
  Motor1.Write(true);
  pwm2.SetDutyCycle(30);
  Motor2.Write(true);
  Thread.Sleep(5000);
  pwm1.SetDutyCycle(30);
  Motor1.Write(false);
  pwm2.SetDutyCycle(30);
  Motor2.Write(false);
  Thread.Sleep(5000);
}
```

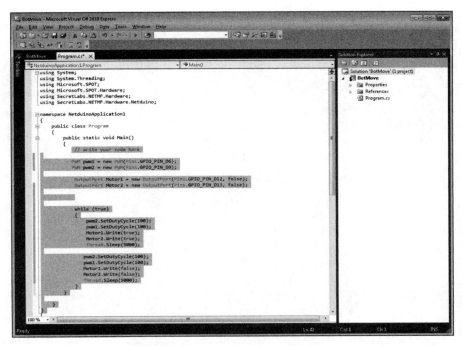

*Add code for controlling your robot's speed.*

You have just told your robot to go forward for 5,000 milliseconds (ms) with the speed reduced by about two thirds by setting the pulse width modulation (PWM) with the following lines:

- **PWM pwm1 = new PWM(Pins.GPIO_PIN_D6);**
- **PWM pwm2 = new PWM(Pins.GPIO_PIN_D9);**

You have set the speed with the following lines:

- **pwm2.SetDutyCycle(30);**
- **pwm1.SetDutyCycle(30);**

The 30 in parentheses represents a value on the available duty cycle for this motor shield, which has a range of 0 to 255. By reducing the duty cycle, you have reduced the speed of the motor.

*Motor shield with jumpers.*

1. Make sure the power cord is disconnected before you load the programming onto your robot.

2. Connect the Micro USB and press play or F5 to load the program.

3. Disconnect the Micro USB cable.

4. Reconnect the power cord and let it run.

Your robot should run forward for 500 ms, then reverse for 500 ms repeatedly at a fraction of the speed from the first program.

## The Least You Need to Know

- You can get creative when mounting your robot. You need wheels and a solid base, but the design is up to you.
- For your robot to be able to move freely, you need to provide power.
- The various I/O pins perform different functions. Make sure you use the proper one for the type of communication you're sending.
- You can use pulse width modulation to control the speed of your motor.

# Sensors

## In This Chapter

- Understanding electronic sensors
- Measuring movement with accelerometers
- Giving direction with magnetometers
- Sensing light, color, sound, gas, and position

A sensor is a device that sends back a signal or indication to represent a measurement. Examples of nonelectronic sensors are a mercury thermometer, which represents temperature displayed by the volume occupied by the mercury in a tube marked with different temperature values, and a weathervane, which rotates to indicate wind direction.

In this chapter we introduce some of the many types of sensors available. With sensors, outside information can be interpreted by electronic circuits. With the addition of instructions given by the designer, that outside information can be acted on.

## What Is a Sensor?

A true sensor must be sensitive to what it is measuring. It must not react to other properties that it is not measuring, nor should it affect the value that it is sensing.

Of course, most sensors will break down in extreme operating conditions. For example, mercury will freeze when temperatures fall below −40°C or boil at temperatures near 360°C. Accurate sensing with a mercury thermometer becomes impractical if not impossible outside of these ranges. Electronic sensors have similar restraints; they must be designed so that under a specified set of operating conditions, the sensors are shielded from interfering signals. For example, if you are measuring the temperature of the air in the room, you wouldn't want the sensor to be shielded from the heat of the circuit or any motors.

Most sensors are related in a linear ratio, meaning that an increase in what is being measured, or sensed, will yield a correlating and linearly related increase in signal. If a microphone senses louder and louder noises, it will register a signal that correspondingly gets stronger and stronger.

The term *transducer* is often used to describe sensors. A transducer represents one type of inputted energy as an output of another type of energy. For example, pressure sensors take physical energy (inputted energy) and convert them to an electrical signal (outputted energy).

# Electronic Sensors

Sensors in electronics can detect and measure many different values: motion (control, location, gravity, speed, and position), the amount of a particular substance in the environment (gases), light (including color), infrared, radiation, sonar, collision/touch, temperature, humidity, and sound or radio frequencies. The wide range of modern, affordable sensors combined with a microcontroller means that you are limited only by your imagination when constructing electronic projects.

Some of the better-documented hobbyist projects using sensors and microcontrollers involve robots that can move autonomously based on sensors that read their environments or radio-controlled vehicles, such as cars or airplanes, that rely on radio signals to control speed and direction. Other popular fields include security systems for home or office that rely on motion sensors and switches, as well as other home automation projects.

Don't be limited by these ideas, though. Imagine building your own breathalyzer interface for your laptop to prevent you from posting on social media websites after drinking too much; or perhaps a magnetic lock and actuator that opens your back door when your dog barks (matched to the voiceprint of your dog's particular bark). Let's look at a few of the more popular and readily available sensors that can be used with microcontrollers. This brief list is by no means exhaustive.

# Accelerometers

An accelerometer measures its orientation relative to either one or more axes (the plural of *axis*, pronounced *AK-sees*). It makes that measurement by measuring acceleration expressed in g-forces (gravitational forces). G-force is the measure of acceleration relative to free fall.

The standard gravity unit of 1 g is the amount of acceleration due to gravity at sea level on Earth. The more planes of movement on an axis (up/down, right/left, or diagonally) that are measured, the more accurate the device will be at tracking motion.

The basic model of an accelerometer can be constructed by a mass connected to a spring. Reorientation (movement in any direction such as height, to the left or right, or angled in any way) from the neutral position will stretch or compress a spring. The displacement of energy on that spring can be measured as a voltage. Think of a gyroscope: any tipping of the gyroscope would reorient it in space.

One of the most common types of accelerometer relies on principles of piezoelectricity. Piezoelectric sensors are created from materials (such as quartz and tourmaline) that produce an electric charge when subjected to pressure due to the rearrangement of their crystalline structures. *Piezo* is the Greek word for pressure.

Modern accelerometers and many other sensors are microelectromechanical systems (MEMS). These are complete systems but at a very small scale, ranging from 20 μm to 1 mm. The small scale and power requirements have enabled their use in a multitude of small electronics. Smart phones use MEMS to orient their image display based on how the phone is held. Tablet computers, digital cameras, and game controllers use accelerometers to enable more realistic game play. Digital cameras and video recorders use MEMS to autocorrect motion in video or still images. The uses are limitless.

# Digital Compasses or Magnetometers

Digital compasses (also known as magnetometers) can give a sensor's orientation relative to magnetic north. The output is generally given in degrees; for example, 90° is east, 180° is south, and 270° is west.

One of the most popular types of magnetometer available for hobbyists relies on the Hall effect. The Hall effect was discovered by American physicist Edwin Hall. Basically, when a conductor or semiconductor with a current running through it is exposed to a magnetic field perpendicular to it, a voltage (called a Hall voltage) is created. The voltage varies proportionally to the strength of the magnetic field, so by measuring the voltage, the sensor can determine magnetic orientation.

**TITANS OF ELECTRONICS**

Edwin Hall (1855–1938) was an American physicist. His graduate work at Johns Hopkins led to the discovery of the effect that bears his name. He made his observations with gold leaf on a glass plate exposed to magnetic fields. He became a physics professor at Harvard and did further research in thermodynamics.

# Light and Color Sensors

Light and color sensors detect light across the spectrum from infrared to ultraviolet and even invisible light energy such as gamma rays and x-rays. Some sensors are simple switches indicating the presence of light, whereas others can distinguish the light intensity and color.

Color is just a different frequency of light wave. There is no physical property of "greenness" or "orangeness"; it is just a matter of frequency and wavelength. The following chart shows the electromagnetic spectrum, which includes the portion of the spectrum that includes visible light.

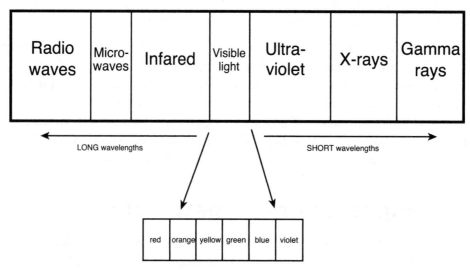

*The light spectrum; visible light is just a portion.*

Light and color sensors are very flexible tools. They can be used in navigation, signal reception, and energy detection. The sensors can be constructed from many technologies, including photoresistors, photovoltaic cells, photodiodes, phototransistors, and charge-coupled devices.

# Microphones

A microphone is a type of sensor as well. It is a transducer that detects sound waves and converts them to an electrical signal. Microphones are essential components in sound recording, telecommunications, hearing aids, voice recognition, and many nonvoice sound applications, including those used in the field of ultrasonic imaging in industrial applications.

Microphone sensors use many different technologies, including condensers (also called a capacitor microphone), piezoelectricity, and fiber optics (by sensing the differences in light intensity bounced off of a reflective diaphragm sensitive to sound vibrations). Microphones designed to work with microcontrollers usually are MEMS and are sometimes referred to as microphone chips.

# Alcohol and Other Environmental Gas Sensors

From sensing toxic chemicals such as carbon monoxide, to determining levels of potentially *combustible* fumes, to detecting alcohol by a breathalyzer, electronic gas sensors have many uses. Gas sensors are generally classified into two types: combustible and toxic and/or noncombustible.

The technologies to detect combustible gases are *catalytic* sensing and infrared (IR) sensors. In catalytic sensing, as the gas comes into contact with a catalytic coil, it oxidizes and the resulting change in resistance on the coil can be measured with the use of a bridge circuit.

**DEFINITION**

To be **combustible** is to be capable of burning or igniting.

To be **catalytic** is to be capable of causing or accelerating a chemical reaction. In this particular case, oxidation is the chemical reaction that results.

IR sensors measure the difference in an infrared beam as it travels through the gas. The difference in the light intensity can be related to the presence of a gas.

For detecting toxic or noncombustible gases, electrochemical gas sensors are a common technology. They use an electrolyte and a porous membrane connected to electrodes. When the gas passes through the membrane, it oxidizes and a current is produced.

# GPS Sensors

The type and specificity of data available via global positioning system (GPS) satellites are incredibly valuable in electronics. GPS can provide location in longitude, latitude, altitude, and time of day. Using that information sampled over time, you can detect speed as well. GPS signals can be integrated into mapping systems, which can be used to provide information, make decisions based on set instructions, and make adjustments in navigation.

The number of projects that can take advantage of GPS sensors is again limited only by your imagination.

## The Least You Need to Know

- Sensors measure a value and transmit a readable signal. Electronic sensors provide an electric signal to describe a value.
- Accelerometers measure speed across multiple axes. This information can be used to give the sensor's orientation in space.
- Digital compasses or magnetometers give direction relative to magnetic fields.
- Light and color sensing relies on reading the wavelengths and frequencies of the light spectrum.
- Microphones can sense sound by converting sound waves to electrical signals.
- GPS sensors receive satellites' information, such as latitude, longitude, and time, which can be integrated with maps and measure speed.

# Electronic Communication

## In This Chapter

- Transmitting and receiving basics
- Understanding various types of signals
- Creating and detecting signals
- Converting visual images to signals

Up to this point you have explored how electronic devices transmit information via electrical signals in a closed system of an electric circuit. This chapter focuses on how electronics can be used to transmit and receive signals across a room, a field, a city, a continent, an ocean, and even into space.

Signal transmission is a very broad area of study. The aim of this chapter is to get you acquainted with the very basics of the topic. At the end of the chapter you will build an FM transmitter to get some hands-on experience with signal transmission.

## The Basics of Electronic Communication

Let's first look at the basic steps involved in using electronics to communicate.

The information being communicated has to be converted into a form that can be transmitted using electricity. That could be a voltage level (amplitude) or another aspect of an electrical waveform. The information, when communicated through electromagnetic methods, is called a *signal*.

A **signal** is information communicated through electromagnetic methods.

Next, the signal needs to be transmitted. This can be done over a dedicated direct connection— such as a copper wire or fiber-optic line—or through some sort of broadcast—such as a Wi-Fi signal over radio waves—or narrowcast—such as infrared waves from your television's remote control device.

An infinite number of signals are constantly being transmitted from an infinite number of sources. Every time you cough, you create a sound wave and a change in the air pressure around you; in other words, you create and send a signal. Light, heat, and UV rays are created by the reactions on the surface of the sun and transmitted in every direction. Everything gives off signals.

For the next step in the communication process, the signal needs to be received. The receiver has to be able to filter out which signals to ignore and which to receive. The more accurate the filter, the less noise, or unwanted signals, is received along with the signal. The amount of signal compared to the amount of noise in a method of communication is called the *signal-to-noise ratio*. The ideal is a high signal, low noise ratio.

The final step in electronics communication is decoding, which is the process of interpreting the signal into a user-friendly communication. The result can be the pictures on your television, an e-mail in your inbox, or the sound of a voice on your telephone.

# The Electromagnetic Spectrum, Revisited

Let's revisit the electromagnetic spectrum from Chapter 21. You probably recall that the key parts of a waveform are the wavelength and the amplitude.

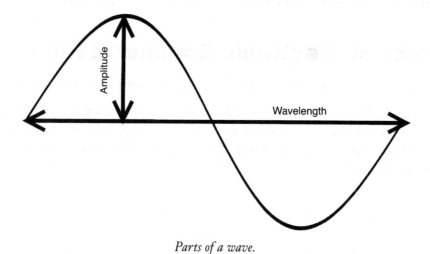

*Parts of a wave.*

The wavelength is the distance over which the wave's shape repeats along an axis. A symbol used for wavelength most often is the Greek letter lambda ($\lambda$). The amplitude is the measurement of the distance either above or below an axis at its greatest point. The frequency of a wave is determined by the velocity (v) of the wave, which is the distance traveled over a period of time. Frequency is expressed in Hertz (Hz), which is the number of waves that pass in a second, and is symbolized as $f$.

The formula showing the relationship between wavelength, frequency, and velocity is $f = v\,/\,\lambda$ where v = velocity of the wave (meters/second, m/s), f = frequency (in Hertz, Hz), and $\lambda$ = wavelength (in meters, m).

For electronic communication, the size of the wavelength can range from $1 \times 10^3$m in the lower end of the spectrum (radio waves) to $1 \times 10^{-11}$m at the high end (gamma rays). The frequency increases as the wavelength decreases, so the very long wavelengths of radio relate to very low frequencies (from 148.5 kHz of long wave AM radio), and the very short wavelengths of gamma rays relate to very high frequencies (with Hz values up to $10^{20}$).

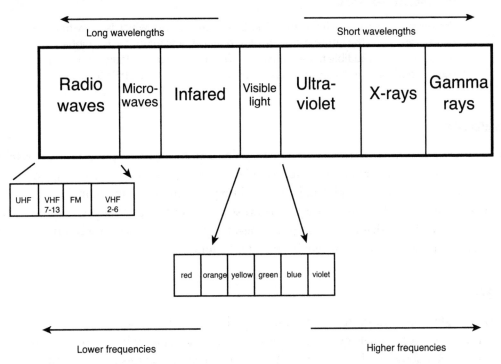

*Longer wavelengths relate to lower frequencies. Frequency increases as we move to the right of this chart.*

The amplitude of a wave is also a key concept in electronic communication. Remember that amplitude can be measured in different ways depending on the type of wave shape. On non-sinusoidal waves, the peak amplitude of a wave is the absolute value of the wave (the highest or lowest point). The amplitude of an electromagnetic wave is a representation of its voltage.

Phase is the synchronous relationship between two waves; it is how closely timed, or synchronized, they are. A wave's phase can't be described without the presence of another wave. When two waves are not synchronized, they are considered to be out of phase. The difference in timing between two waves' movements along an axis determines the amount they are out of phase.

## Radio Waves

Radio waves are the most widely used signals in electronics communication. They are used in radio and television broadcasting, mobile phones, and wireless computer networking signals such as Wi-Fi and Bluetooth. The range of radio wave frequencies encompasses ultra-low frequency signals to high-frequency signals, each with its own properties of transmission. Low frequencies can travel long distances and aren't dependent on line of sight between transmission points. High frequencies are much more dependent on line of sight transmission and are less likely to be able to transmit through buildings or natural obstacles.

> **WATTAGE TO THE WISE**
>
> The robot project in this book uses 42 kHz frequency waves, which is a low-frequency radio band. It is in the ultrasonic range of frequencies, meaning that is a higher frequency than is audible to the human ear. The human ear can detect and interpret as sound frequencies in the acoustic range, which goes from from 20 Hz to 20 kHz.

## Microwaves

Microwaves are used in short-range point-to-point communication applications, in satellite communications, and in radar applications. The advantages of using microwaves for communication are that they have a high bandwidth and they use line-of-sight technology, so the same frequency can be used as long as they don't cross paths. The disadvantages of using microwaves for communication is that they do require line-of-sight relationships between transmitter and receiver and they are affected by environmental factors.

## Infrared

Infrared light is used in short-range communication applications such as in sensors or between two nearby devices. Signals can be sent between LEDs and photodiodes, or with infrared lasers across longer distances. Many TV and other remote control devices use infrared light.

## Visible Light

The visible light portion of the spectrum can be used to transmit and receive information, but it is not widely used for this purpose. However, there is considerable research into its use as a very large bandwidth transmission medium.

## Ultraviolet, X-Rays, and Gamma Rays

Using the high frequency end of the electromagnetic spectrum as a transmission signal is problematic because the *radiation* associated with this portion of the spectrum can be harmful to health. High-frequency waves do play a role in many types of sensors; and photodiodes measure these waves to detect many naturally occurring signals.

**DEFINITION**

**Radiation,** scientifically speaking, refers to all energy emitted in rays or waves.

Higher-frequency waves are known as **ionizing radiation.** These waves are energetic enough to separate electrons from atoms, creating ions. That doesn't necessarily mean that they are dangerous to health, but this effect does mean that these rays should be shielded to reduce the possibility of damage to the DNA in cells and as such, their use as a means of electronic communication has to be balanced against their potential harm.

# Encoding and Decoding a Signal

A signal in electronics communication can be as simple as a measurement of voltage or amplitude. For example, for a pressure sensor that turns on a circuit to provide a certain threshold voltage, the only signal required is enough pressure. This type of electronic communication requires very little sophistication in encoding or reading the signal.

Other methods of electronic communications involve modulation. Modulation starts with what's called a *carrier signal wave.* The carrier signal wave is transmitted as a constant signal wave layered with an information signal wave. Modulation is the process of encoding the information that is sent along with the carrier signal.

There are three primary methods of modulation: amplitude modulation (AM), frequency modulation (FM), and phase modulation (PM). Both analog and digital communications use these methods to send signals.

## Amplitude Modulation

Amplitude modulation encodes the information signal wave within the amplitude of the carrier signal wave. In the accompanying figure, notice how the carrier wave is the outside limits of the wave. The information is stored within the "envelope" of the carrier wave.

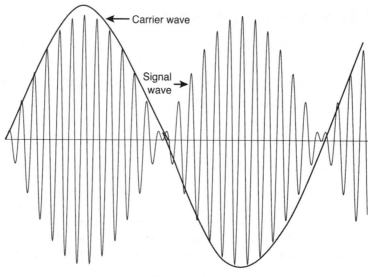

*Amplitude modulation.*

## Frequency Modulation

With frequency modulation, the carrier signal wave is set to a particular steady frequency and the information is set within minor variations above and below that frequency. It is a bit harder to visualize, but note in the following figure that the information is encoded above and below the carrier signal wave.

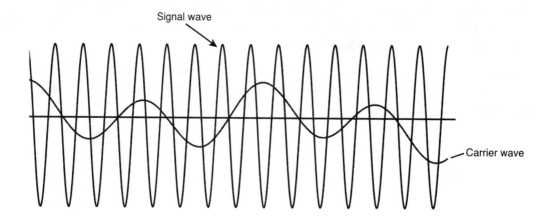

*Frequency modulation.*

# Phase Modulation

In phase modulation, the encoding and the decoding of the information signal wave requires a timing key. The information signal wave is set off-phase from the carrier signal wave. The difference in phase contains the information. The timing key is required at the transmitting end and the receiving end of the communication.

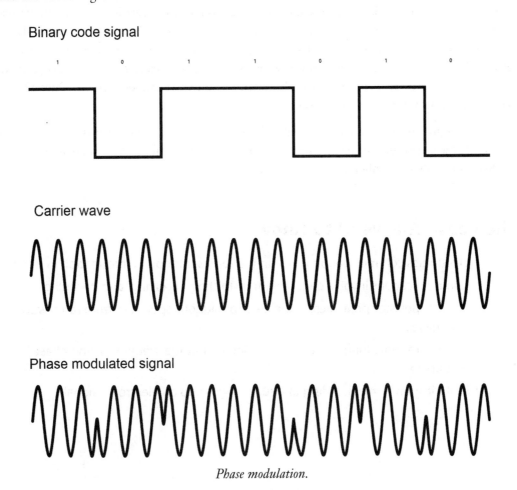

*Phase modulation.*

The receiver uses a type of filter called a *bandpass filter* to select only signals that fall within a certain range of frequencies or wavelengths. The information is then separated from the carrier wave signal by a demodulator.

**WATTAGE TO THE WISE**

A modem is a device that both modulates and demodulates. It is a combination of the two terms *modulator* and *demodulator*. The communication of data across the Internet requires both the sending and receiving of signals, and modems serve as the gateways for this communication.

A demodulator may also be called a *detector* even though it does more than detect. It decodes the signal by separating the carrier signal from the information.

# Rasterization

To send a visual image via electronic means it's necessary to convert it into a mathematical representation of a visual image, a process called *rasterization*. A three-dimensional (3D) visual image is described geometrically as a series of polygons composed of various triangles. The relationship between these shapes can be expressed mathematically to create a two-dimensional (2D) image that can be displayed on a television, computer, or camera screen. Once the image is expressed mathematically, it can be sent as an electronic signal.

When the signal for an image is received, it is constructed of a series of instructions for pixels or dots that are arranged according to the rasterized signal. Information such as color and brightness can also be embedded in the signal.

## The Least You Need to Know

- Electronics communication relies on the representation of information in waveform.
- Communications involve encoding, transmitting, receiving, and decoding.
- Signal types are spread across the electromagnetic spectrum from radio waves to gamma rays.
- Signals are sent along carrier waves using amplitude, frequency, and phase modulation.
- Rasterization is the encoding of visual images into an electronic signal.

# Lab 22.1: Building an FM Stereo Transmitter

In this lab you will build an FM stereo transmitter from a kit manufactured by Electronic Rainbow and sold by Jameco Electronics at www.jameco.com (manufacturer no. FMST-100). Included in the package are instructions, the schematic, and all of the parts necessary for construction.

An FM stereo transmitter enables you to send an audio signal from a source to an FM receiver. For example, you can send an audio signal from a stereo in your living room to speakers throughout the whole house. This project tunes from 88 MHz to 108 MHz, which is the FM frequency band in most countries.

The radio signal is created with the use of a current applied to an oscillator, in this case a 38.0 KHz crystal. This crystal is a piezoelectric material that creates its own electrical signal at very precise frequencies. You can tune to a specific frequency by adjusting the level of voltage applied through the use of capacitors and resistors in an oscillator circuit.

The kit includes a printed circuit board (PCB) with clear markings that indicate where each component should be placed as well as the appropriate polarity direction of the components.

**Materials:**

>      1 FMST-100 kit
>
>      1 antenna (you can use any RCA cable)
>
>      1 9 V battery

**Instructions:**

The instructions that come with the kit give good step-by-step advice, but here's an overview of the process:

1. Place and solder the resistors as indicated by the kit instructions.

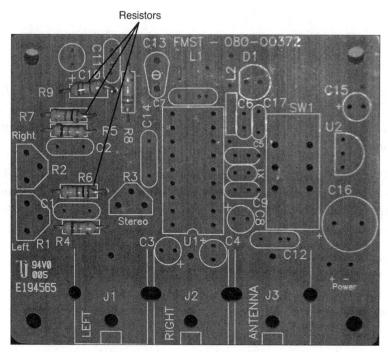

*Solder the resistors to the PCB.*

2. Place the IC socket, the inductor, and the crystal as indicated on the kit instructions.

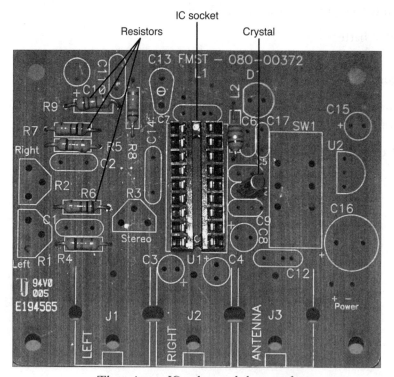

*The resistors, IC socket, and the crystal.*

3. Place the capacitors as indicated on the kit instructions. Remember that on a capacitor, if the polarity isn't marked clearly, the long lead side is positive.

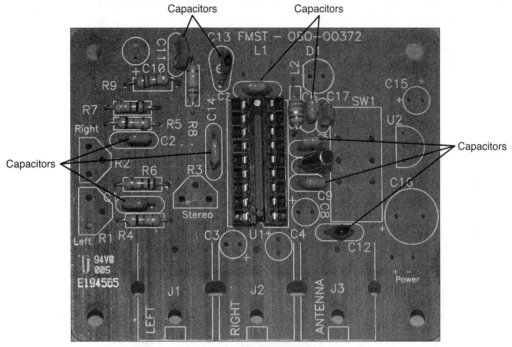

*Add the capacitors.*

4. Place the remaining miscellaneous parts, being sure to carefully look over all of your solders and double-check placement of all parts before inserting your IC.

5. Add an antenna.

*The completed project. Note the RCA cable attached on the right.*

Follow the instructions for testing and tuning your transmitter. To tune your transmitter, use the plastic tool (do not use a metal tool!) included in the kit. For coarse tuning, adjust the coil marked L1. To make fine frequency adjustments, also called fine tuning, adjust the coil marked R3.

You can adjust the volume (amplification) in each channel (right and left) by adjusting R1 for the right channel and R2 for the left channel.

*Note the adjustment dials (potentiometers) in the front.*

# Adding Sensors
# to Your Robot

## In This Chapter

- Attaching an ultrasonic range finder to your robot
- Setting up the power switch
- Coding for collision avoidance
- Letting your robot go

Once the wheels are spinning under the control of your program, you have successfully built a robot. But you can do so much more with a combination of new programming and add-on sensors. This chapter gets you going by adding a sensor that can detect objects and their distance through the use of ultrasound waves.

## Adding Collision Control

Your robot goes backward and forward, which is cool enough, but let's take it further. Let's have it go forward until it senses an object. Then we can have it turn and go the other way until it encounters something again. This process is known as *collision control*.

You will be using the MaxBotix Ultrasonic Range Finder you ordered back in Chapter 18 to tell the bot when it's within a certain distance from an object. You will use that information combined with instructions written in C# code to perform these actions.

# The Ultrasonic Range Finder

First you need to understand what the range finder does and how it works. We are using an LV-EZ1 Ultrasonic Range Finder from MaxBotix, Inc. Download the data sheet for the range finder from MaxBotix's website at www.maxbotix.com.

*The LV-EZ1 Ultrasonic Range Finder by MaxBotix.*

The range finder works by sending out a 42 kHz ultrasonic sound wave and then calculates the time it takes for the sound wave to be reflected back based on the speed of sound. Connect the range finder to your Netduino microcontroller using the analog output pin labeled AN on the sensor.

For every inch it measures, it produces a signal to the microcontroller in increments of 10 mV from 0 to 255 inches. So if you are 10 inches away from an object, the signal would be 100 mV. By sensing the signal you can calculate the distance and give instructions accordingly. You can view this action by using your breadboard and a DMM.

## Adding the Sensor to Your Robot

You are now ready to attach the MaxBotix Ultrasonic Range Finder MaxBotix LV-EZ1 sensor to your robot. Grab your robot and let's get to work.

**Materials:**

Your robot

MaxBotix Ultrasonic Range Finder MaxBotix LV-EZ1 sensor

Small breadboard

Jumper wire

100 μF capacitor

100 Ω resistor

## Instructions

1. Use double-sided tape to secure the small breadboard to the front of the robot.

2. Use the following diagram to make connections between the motor shield on your robot, the breadboard, and the sensor. Add the sensor to the breadboard so that the sensing side is facing forward (the circuit board will be facing the rear). The sensor will insert directly onto the breadboard. The connections between the breadboard and the motor shield should be made with jumper wire. Notice the placement of the 100 μF capacitor and the 100 Ω resistor.

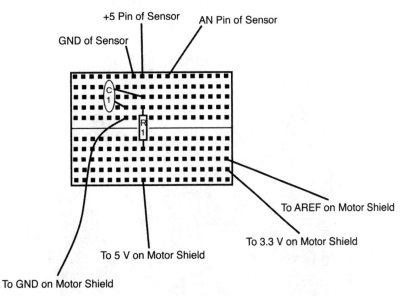

*Connection diagram for sensor; connections to be made on the small breadboard.*

*Completed robot side view.*

*Completed robot top view.*

# Adding a Power Switch

To make it easy to power your robot on and off, it's time to connect a single-pole, single-throw (SPST) switch between the battery and the power input on the Netduino microcontroller. Here are the steps:

1. On the 9 V to Barrel Jack Adapter cable, working from the end that has the battery connector, strip off approximately 6 inches of the outer insulation, exposing the red and black wires.

2. Determine where you want to place your switch, making sure to leave plenty of ground clearance between your robot's base and the running surface.

3. Attach your switch to your base. Bring the black wire to the switch leads to see where to make your cut.

4. Cut the black wire and strip off enough of the black insulation to make your connections.

5. Solder the wires to the switch lead.

# Planning and Writing the Code

Let's first plan out what we want to do and create a flowchart of what we want to happen.

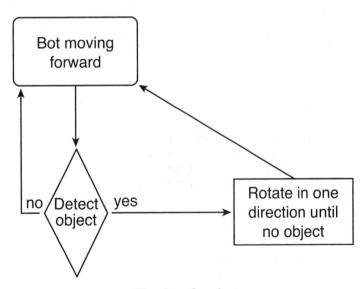

*Flowchart for robot.*

This flowchart may seem very simple, but as you work on more complicated programs, you will find that using a flowchart helps to produce bug-resistant, if not bug-proof, programs.

You can use proper flowchart symbols if you wish, but even a list of steps will help you accomplish your goals.

Now we can add code to provide collision avoidance. Insert this code just below the comment line "//write your code here" but above the PWM lines:

```
AnalogInput a5 = new AnalogInput(Pins.GPIO_PIN_A5);
```

This code creates a new analog input named a5 and assigns it to analog pin 5. Now, add this code below the last output motor line:

```
int i = a5.Read();
        if (i > 20)
```

This will read the input and, if the value is greater than 20, the robot will move forward. If the value is less than or equal to 20, the "else" statement will kick in. This reverses one of the motors, which will make one wheel reverse and cause the robot to spin away from the object until the input value is once again above 20. When it reaches 20, it will return to normal forward motion.

Here's the full code for your robot so far:

```
// write your code here
        AnalogInput a5 = new AnalogInput(Pins.GPIO_PIN_A5);

        PWM pwm1 = new PWM(Pins.GPIO_PIN_D6);
        PWM pwm2 = new PWM(Pins.GPIO_PIN_D9);

        OutputPort Motor1 = new OutputPort(Pins.GPIO_PIN_D12, false);
        OutputPort Motor2 = new OutputPort(Pins.GPIO_PIN_D13, false);

        int i = a5.Read();
        if (i > 20)
            while (true)
            {
                pwm2.SetDutyCycle(100);
                pwm1.SetDutyCycle(100);
                Motor1.Write(true);
                Motor2.Write(true);
            }
        else
            {
              pwm2.SetDutyCycle(100);
              pwm1.SetDutyCycle(100);
              Motor1.Write(false);
              Motor2.Write(true);
            }
            }
        }
    }
```

# Letting Your Robot Roam

Now it's time to load the programming to your computer and let your robot roam free in the world. First, make sure the power switch is off on the robot. Connect the Micro universal serial bus (USB) cable from your computer to your Netduino and press the **F5** key to load the code onto the robot. After it is done loading, disconnect the USB cable, place the robot on the floor, flip the power switch, and watch your robot go!

## The Least You Need to Know

- You can attach a sensor to your robot and program it to create a collision avoidance system.
- The MaxBotix Ultrasonic Range Finder is a sensor that uses sound waves to detect objects.
- You can add a power switch to turn your robot on and off.
- In writing the code, you need to decide which actions should occur with the data you sense.

## Lab 23.1: Sensing Distance

In this lab you will see how the sensor detects distance from an object and the output it supplies as you read it on the DMM.

**Materials:**

MaxBotix Ultrasonic Range Finder LV-EZ1

Breadboard

5 V power supply

Jumper wire

DMM

**Instructions:**

1. Connect the 5 V power supply to the pin labeled +5.

2. Connect the pin labeled GND to GND.

3. Set your DMM to read **VDC**.

4. Connect the black lead to the GND.

5. Connect the red lead to the pin labeled AN.

6. Turn the power on.

7. Move your hand in front of the sensor; as you do, the voltage will change. The closer you are, the lower the voltage produced; the farther you move away, the higher the voltage produced.

*MaxBotix Ultrasonic Range Finder LV-EZ1 with pins labeled.*

**accelerometer**   A sensor that measures acceleration relative to free fall.

**actuation force**   The force required to throw the actuator.

**actuator**   A mechanical system that controls a device or system.

**algorithm**   A mathematical formula.

**alternating current (AC)**   Electron flow in both directions alternating between forward and backward. It also has changes in polarity.

**alternator**   An AC electrical generator.

**ampere**   A coulomb of current that moves through a point in one second.

**amplifier**   A circuit or device that increases a signal.

**amplitude**   The measurement of the distance of any point of the wave that is above or below the center or mean line.

**analog**   Describes systems that are continuous as opposed to stepped. An analog signal can be represented by a wave and not individual, unconnected points.

**anode**   An electrode that receives current.

**armature**   The rotating arm in a brushed motor that interacts with the stator to create a torque on it.

**assembly language**   A low-level programming instruction that uses symbols to represent the machine code.

**atom**   The smallest unit of each element.

**battery**   A power source that uses chemical energy to produce and store electrical energy.

**binary system**    A base-2 numbering system, with two digits: 1 and 0.

**binary-coded decimals**    A method of depicting decimal values by assigning a binary value to each of the decimal digits.

**bit**    A binary digit.

**Boolean algebra**    A system that applies classical logic to mathematical operations.

**breadboard**    A reusuable solderless circuit board.

**brushed motor**    A type of motor that uses brushes as contacts.

**bus**    In electronics, the electrical connection between components; in computing, the transfer system of information between components.

**byte**    Eight bits.

**capacitance**    The ability to store charge.

**capacitors**    An electrical component that can store charge (electrons).

**catalytic**    Capable of causing or accelerating a chemical reaction.

**cathode**    An electrode that current flows out of.

**central processing unit (CPU)**    An element of a computing system that performs instructions.

**checksum**    A method of error protection that compares transmissions to a stored algorithm.

**chip**    In electronics, an integrated circuit that is embedded on a material.

**circuit**    A pathway that allows current to flow through components.

**circuit breaker**    A reusuable component that protects against short circuits.

**circuit diagram**    A pictoral description of an electronic circuit.

**combustibility**    The capability to burn or ignite.

**commutator**    A ring with gaps that makes the connections between the power brushes and the armature in a brushed DC motor.

**component**    An element in an electronic circuit; they can be passive or active.

**conductivity**    The tendency of a material to allow the free flow of electrons.

**conductor**   Material that allows the flow of electrons.

**contact bounce**   The energy sent back into a circuit from the mechanical force of contact being made.

**conventional theory**   The early belief that current travels from positive to negative.

**coulomb**   A unit of measurement equal to approximately 6.25 × 1018 electrons.

**crystalline structure**   Structure in solids that have an orderly, repeating pattern.

**cycle**   One complete evolution of a wave's shape.

**Darlington pairs**   Two transistors sold as one with the leads arranged as if they were one unit.

**data sheet**   The instructions, schematics, and other technical information provided by a component, circuit, or the device's manufacturer.

**daughterboard**   A circuit board that connects directly to the central processing unit or motherboard.

**decoder**   A circuit that decodes a signal from machine-readable code.

**desoldering**   The process of removing solder.

**dialect**   A form of programming language specific to a particular programming environment.

**digital**   Systems that are composed of discrete steps. A digital signal is not a continuous wave, but a series of unconnected points.

**diodes**   Components that permit electric flow in a single direction and act against flow in the opposite direction.

**direct current (DC)**   Electron flow that is unidirectional.

**donor impurity**   A material that when added to a semiconductor causes it to lose or gain electrons.

**doping**   In electronics, adding impurities, usually to a semiconductor.

**electric arc**   The transfer of current through a gas, not through a prescribed circuit or conductive material but instead by creating a plasma. Lightning is an example of an electric arc.

**electricity**   The energy created by the movement of charged particles.

**electrode**   A terminal or connector between a conductor and a nonmetallic part of a circuit.

**electrolyte**   A substance that has free ions that can become conductive.

**electromagnetic force**   The attractive force that exists between positively charged protons and negatively charged electrons.

**electromotive force**   The "push" that gets the electrons jumping from one atom to another, sending a current of electrical flow along the way. *See also* potential difference, voltage.

**electron theory**   Current travels from negative to positive. This was opposed to conventional theory, which was the misunderstanding held by early experimenters who mistakenly thought that current traveled from positive to negative.

**electronics**   The study of electron flow (electricity) and the devices that control electron flow to perform work.

**electrostatic force**   The force that bonds the protons and the electrons of an atom.

**element**   The most basic unit of each distinct type of matter.

**encoder**   A circuit that represents a signal in machine-readable code.

**engineering notation**   Scientific notation that uses powers of 10 that are divisible by 3 to coordinate with metric prefixes.

**EPROM**   Electrically programmable read-only memory.

**farad**   A unit of measurement equal to the charge in coulombs required to raise the voltage across the capacitor by one volt.

**flux**   A chemical used to clean oxidation from components prior to soldering.

**forward error correction**   A method of error protection that sends redundant transmissions.

**frequency**   The number of complete cycles in a given amount of time.

**function generator**   An instrument that creates waveforms and signals.

**fuse**   A single-use component that protects against short circuits.

**gain**   The increase in a signal.

**gravitational force**   The measure of acceleration relative to free-fall.

**heat sink**    A device that assists in cooling by increasing the surface area.

**Hertz**    Equal to one cycle per second.

**hexadecimal**    A base-16 numbering system; it has the digits 0–9, plus A–F.

**induction**    The production of voltage as a conductive material travels through a magnetic field.

**insulators**    Materials that resist the flow of electrons.

**integrated circuit (IC)**    A miniaturized circuit that rests on a semiconductor base, also known as a chip.

**inverter**    A circuit or device that converts DC to AC.

**Joule's First Law**    Derived from Ohm's Law, the formula that relates power, voltage, and current: $P = V \times I$.

**logic gate**    A circuit that performs a logical function.

**logic probe**    A device that indicates the binary state of a circuit.

**machine code/language**    Programming instructions understandable by the central processing unit of a computer.

**magnetometer**    A sensor, also known as a digital compass, that senses orientation relative to magnetic north.

**matter**    All physical objects; everything that has mass (measurable stuff) and volume (measurable occupation of space) is matter.

**mechatronics**    A scientific field that synthesizes mechanical and electrical engineering, electronics, and control theory.

**microcontroller (MCU or $\mu$C)**    A self-contained embedded computer.

**microphone**    A transducer, or sensor, that detects sound waves and converts them to an electrical signal.

**microprocessor**    An integrated circuit that serves as the CPU for a computer.

**mixed signal**    A circuit or device that uses analog and digital signals.

**n-type semiconductor**    A semiconductor that has been doped to have an excess of negative charge.

**neutron**    A subatomic particle that doesn't carry an electrical charge.

**nibble**   Four bits.

**nonvolatile memory**   Memory that is stored even when power is not supplied.

**nucleus**   The central part of an atom that contains the protons and neutrons.

**Ohm**   The level of resistance that allows one volt of electromagnetic force to move one ampere across two points on a circuit.

**Ohm's Law**   The basic formula relating voltage, current, and resistance: $V = I \times R$.

**operand**   A quantity that has a mathematical or logical operation performed on it.

**oscilloscope**   A tool that depicts the waveforms produced by electrical current, both AC and DC. It allows you to analyze quickly the voltage and wave characteristics of the current.

**oxidation**   A chemical reaction between a material and oxygen; or the process of an atom losing electrons.

**p-type semiconductor**   A semiconductor that has been doped to have an excess of positive charge.

**parity bit**   A bit that is used to check for errors in transmission that relies on whether the value is odd or even.

**period**   The amount of time to complete one cycle.

**periodic table of elements**   A pictorial description of the known elements.

**phase**   A synchronous relationship between two waves.

**phase shift**   Adjustment to the relationship between two waves.

**photodiode**   A diode that converts light energy into electrical current.

**photoresistor**   A resistor that is reactive to light energy.

**piezoelectricity**   Electric charge created in certain materials (usually with a crystalline structure) when pressure is applied.

**polarity**   An electrical charge (negative or positive).

**pole**   A setting on a switch.

**potential difference**   The "push" that gets the electrons jumping from one atom to another, sending a current of electrical flow along the way. *See also* electromotive force, voltage.

**potentiometer**   A component that acts as a voltage divider to control the amount of signal that flows through a circuit.

**power**   Voltage multiplied by current.

**primary battery**   A battery that is nonrechargeable after depletion of its stored energy.

**proton**   A subatomic particle that carries a positive electrical charge.

**RAM**   Random access memory.

**reading**   In computing, retrieving data.

**rectifier**   A circuit or device that converts AC to DC.

**relay**   A switch that uses an electromagnetic field to control the opening and closing of a circuit.

**resistance**   The tendency of a material to resist the flow of electrons.

**resistors**   Basic electronic components that increase the resistance of a circuit.

**robotics**   The study of robots, their construction, and control.

**ROM**   Read-only memory.

**scientific notation**   A system that represents decimal numbers as a product of the numbers 1 through 9 multiplied by a power of 10.

**secondary battery**   A battery that can be recharged after depletion.

**semiconductors**   Materials, generally with crystalline structures, that are in the middle range of materials as to conductivity or resistance.

**sensor**   A device that feeds back a signal or indication to represent a measurement.

**shell**   The area around the nucleus where the electrons travel.

**solder**   A metal that is melted to form connections between wires and components.

**soldering iron**   A tool used to melt solder to form connections between wires and components.

**source code**   The lowest human-readable programming instructions.

**stator**   A stationary magnet used to provide a magnetic field in a motor.

**switch**   A component or device that opens or closes a circuit.

**thermistor**   A resistor that is reactive to temperature.

**throw**   A contact that completes a circuit.

**tolerance**   The percentage in possible variation from the stated value.

**torque**   Rotational force; a twisting force around an axis or pivot point. It is sometimes referred to in mechanical engineering as moment or moment of force.

**trace**   A wire set into a circuit board.

**transducer**   A device that converts one type of inputted energy into an output of another type of energy.

**transformer**   A device that can either "step up" or "step down" voltage in AC through the operation of mutual induction.

**transistor**   An electronics component that amplifies signals or acts as a switch in a circuit.

**truth table**   A chart of the possible outcomes in a logical decision depending on the inputs.

**universal memory**   The goal of developing memory that is both affordable to produce and energy-efficient. It should be fast, nonvolatile, and resistant to magnetic interference.

**valence band/valence shell**   The outermost electron shell of an atom.

**varistor**   A variable resistor that is designed to protect circuits from excessive voltage.

**volatile memory**   Memory that is lost when power is not supplied.

**voltage**   The "push" that gets the electrons jumping from one atom to another, sending a current of electrical flow along the way. *See also* electromotive force, potential difference.

**voltaic cell**   A chamber that is composed of two half cells that facilitates the chemical reaction that produces electrical energy in a battery.

**watt**   A unit of measurement equal to one volt pushing one amp of current.

**waveform**   A graphical depiction of a signal across the x and y axis to describe its features.

**wetting**   In soldering, the process of reducing the surface tension to ease the flow of solder.

**writing**   In computing, storing data in memory.

# Component Shopping List

When setting up your workspace it's always a good idea to have a basic pantry of supplies. In this appendix we list the variety of components necessary to complete all of the labs in the book (the robot has its own shopping list; see Chapter 18). We do not include the items that are your essential tools of the trade as described in Chapter 4.

You'll need to come up with an organization system for all of your supplies. It can be made of baby jars, a tackle box, or what we use: a set of small plastic drawers. Think small, and be sure to have a system for labeling everything. We use a label maker, but a marker and some masking tape might work for you.

Handle the components with care. Troubleshooting becomes incredibly complicated if you are dealing with damaged parts. Diodes are especially delicate, and we recommend testing them with your DMM before soldering to ensure they are in working order. And be sure to use a heat sink when soldering to protect components.

LM386 operational amplifier (op-amp)

AA battery

9 V battery (you might want several)

9 V battery connector with leads

2 flashlight bulbs (lamps)

Perf board

Masking tape

Roll of 60/40 solder

Roll of lead-free solder

Bottle of liquid soldering flux

¼-inch mono audio jack

8 Ω speaker

0.01 µF capacitor

0.047 µF capacitor

10 µF electrolytic capacitor

100 µF electrolytic capacitor

2 220 µF electrolytic capacitors

7490 decade counter

7448 decoder

Red light-emitting diode (LED)

Common cathode 7-segment LED display

Photocell

5 KΩ potentiometer

10 Ω resistor

2 100 Ω resistors

7 270 Ω resistors

390 Ω resistor

470 Ω resistor

560 Ω resistor

100 KΩ resistor

1 MΩ resistor

25 Ω rheostat

8-pin socket

2N2222 NPN transistor

SPDT switch

2 SPST switches

8 dual-inline-package (DIP) switches

555 timer IC

Kits:

Power Supply Kit 20626 (JE215) from Jameco Electronics

FM Stereo Transmitter Kit (FMST-100) from Jameco Electronics

# Electronics Timeline

**1600**  British scientist William Gilbert first uses the word *electricity*.

**1660**  Otto von Guericke invents the first electrostatic generator.

**1729**  Stephen Gray experiments with the concept of conductivity of electricity.

**1745**  While working at Leiden University, Pieter van Musschenbroek invents the first storage device (a capacitor) for static electricity, the Leyden Jar (sometimes spelled *Leiden*).

**1747**  William Watson creates a circuit that carries the current from a Leyden Jar. He also adds lead foil to the inside of the jar to increase its capacity. Benjamin Franklin discusses his ideas about electrical fluid composed of particles. Henry Cavendish measures the conductivity of different materials.

**1750–1752**  Benjamin Franklin develops the lightning rod; he explains that lightning is electricity.

**1800**  Alessandro Volta invents the first electric battery and shows that electrical current can travel through wires.

**1820**  The electromagnet is invented by D. F. Arago.

**1820–1821**  Charles Babbage proposes a Difference Engine, which is to be a massive steam-powered mechanical calculator. He later proposes the Analytical Engine, which uses punch-cards based on the Jacqard loom.

**1821**  Michael Faraday invents the first electric motor.

**1826**  Georg Simon Ohm introduces his law: $V = IR$.

**1831**  Faraday publishes his principles of induction by electromagnetism, generation, and transmission.

**1842**   Ada Lovelace Byron publishes her analysis of Babbage's ideas and lays the foundations for computer programming theory.

**1854**   George Boole publishes his mathematical ideas, which will form the basis for Boolean algebra.

**1878**   Edison Electric Light Company is founded.

**1879**   The first commercial lighting system is installed in Cleveland, Ohio; it uses arc lighting. Thomas Edison introduces his incandescent lamp.

**1882**   The first hydroelectric power station opens in Wisconsin.

**1886**   William Stanley introduces an alternating current (AC) electric system, and the work of Frank Sprague develops the technology to use a transformer system to make long-distance AC transmission possible. The Westinghouse Electric Company is founded.

**1888**   Nikola Tesla invents the rotating field AC alternator.

**1893**   Tesla addresses the Franklin institute in Philadelphia and explains the theory of radio communication. Westinghouse demonstrates an electricity generation and transmission system at Chicago's World Exposition.

**1896**   Guglielmo Marconi develops the first practical radio system.

**1897**   The electron is described by J. J. Thomson.

**1904**   John Ambrose Fleming invents the thermionic valve, the first radio tube diode.

**1906**   Robert von Lieben and Lee De Forest develop the amplifier tube, the triode.

**1907**   De Forest invents the vacuum tube electric amplifier, the Audion.

**1912**   Ewin Armstrong introduces the regenerative feedback amplifier and oscillator.

**1920**   The first news program is broadcast on radio in Detroit.

**1922**   Entertainment broadcast radio is launched in England.

**1928**   Philo Farnsworth presents the first public demonstration of electronic television.

**1937**   Claude Shannon publishes his paper laying the basis for the application of binary numbers to switches in electronic circuits, which becomes the basis of digital circuit design.

**1940**    Bell Labs demonstrates the Complex Number Calculator, which is able to perform calculations over the telephone line.

**1941**    Konrad Zuse creates the Z3 computer in Berlin. It is destroyed in a 1943 bombing raid. The Bombe computer is developed from a Polish design by the British. It is used by the Allied forces to decrypt Nazi war communications.

**1945**    John Von Neumann outlines the structure of a stored-program computer, the Electronic Discrete Variable Automatic Computer (EDVAC). It introduces the ideas for the binary system and leads the way for digital computing.

**1947**    William Shockley, John Bardeen, and Walter Brattain of Bell Labs invent the transistor. Electronic Numerical Integrator and Computer (EDVAC) is developed at the University of Pennsylvania.

**1948**    Freddie William and his colleagues invent the Random Access Storage Device.

**1949**    Electronic Delay Storage Automatic Calculator (EDSAC) is introduced at Cambridge University. It uses subroutines stored on punched paper tapes and thus is the first practical stored-program computer. It is the realization of the EDVAC idea.

**1950**    Engineering Research Associates sells its ERA 1101 (the first commercially produced computer) to the U.S. Navy. The National Bureau of Standards constructs two large computers (the SEAC and the SWAC) to test components. The Standards Eastern Automatic Computer (SEAC) is the first to use all-diode logic.

**1951**    Another early commercial computer, the Ferranti Mark 1, is released. The UNIVAC I built by Remington Rand is sold to the U.S. Census Bureau.

**1953**    International Business Machines (IBM) releases its model 701, the first electronic computer.

**1954**    Texas Instruments sells the first consumer transistor radio, the Regency TR1. IBM releases the IBM 650 magnetic drum calculator, the first mass-produced computer.

**1956**    The IBM 305 hard disk is released. It can store 5 MB.

**1958**    NEC builds the NEAC 1101, Japan's first electronic computer.

**1959**    A germanium-based integrated circuit is patented by Jack Kilby. Robert Noyce develops a silicon-based integrated circuit. IBM introduces the 7000 series mainframes—transistor–based computers.

**1960**   Massachusetts Institute of Technology (MIT) students write the first computer video game on Digital Equipment Corporation's (DEC) PDP-1.

**1961**   IBM releases the 1401 Data Processing System.

**1962**   The metal-oxide semiconductor field effect transistor (MOSFET) is invented by Steven Hofstein and Frederic Helman. It is cheaper, smaller, and uses less power than earlier transistors.

**1964**   Beginners' All-purpose Symbolic Instruction Code (BASIC) is introduced. IBM releases the System/360 mainframe computer. It introduces the concept of expandability and the use of peripherals.

**1965**   Robert Lucky of Bell Labs invents the automatic equalizer. Gordon Moore publishes the paper that popularizes the idea that integrated circuits will be able to be twice as complex (that is, that they could be composed of twice as many components) year over year for at least 10 years. This comes to be known as Moore's Law.

**1966**   Hewlett-Packard introduces a general-purpose business computer that supports several computer programs, including BASIC and IBM Mathematical Formula Translating System (FORTRAN).

**1968**   Seymour Cray designs the CDC 7600, from CDC, which uses the idea of peripheral processors working together with a central processing unit. Doug Engelbart develops a word processor, a collaborative application, and an early hypertext system.

**1969**   A team at MIT and the Defense Advanced Research Projects Agency (DARPA) creates the Advanced Research Projects Agency Network (ARPANET); it introduces the concept of data transfer through packets accessible to multiple machines. ARPANET laid the basis for future systems that would develop into the Internet. Bell Labs introduces UNIX. Dynamic random access memory is invented by Robert H. Dennard.

**1971**   Ted Hoff and his team at Intel release the first commercial microprocessor, the Intel 4004. Kenbak-1, the first kit-based personal computer, is introduced. Bill Gates and Paul Allen develop a computer traffic-analysis tool.

**1972**   Hewlett-Packard releases the HP-35, a scientific handheld calculator advertised as "a fast, extremely accurate electronic slide rule." Atari releases Pong for the arcade.

**1973**   Intel releases the Intel 8080, used in the MITS Altair 8800.

**1974**  Xerox Palo Alto Research Center (PARC) develops the Alto, which uses a mouse for input and can be linked in a small network. Never sold, it was given to universities for research and development. Bob Kahn and Vinton Cerf introduce a set of ideas that would become the Internet, and their paper includes what is considered the first use of the term *Internet*.

**1975**  The Altair 8800 is released. It is sold by mail order as a kit and uses Altair BASIC (developed by Paul Allen and Bill Gates) as its software language. The Intel 8048 is the first commercially available chip with both random access memory (RAM) and read-only memory (ROM) on the same chip.

**1976**  Steve Wozniak designs the Apple I.

**1977**  The Commodore Personal Electronic Transactor (PET) is introduced. It arrives fully assembled and is easy to operate. The Apple II is released; it can be attached to a color TV to produce color graphics. Radio Shack sells 10,000 TRS-80s in the first year; it is the first computer designed for the computer novice.

**1978**  International Packet Switched Service (IPSS) is launched by the British Post Office, Western Union International, and Tymnet.

**1979**  Usenet is established as a Unix-to-Unix Copy (UUCP) link between the University of North Carolina–Chapel Hill and Duke University. Compuserve becomes the first public e-mail service for personal computer users.

**1980**  Built-in Self Testing (BIST) circuit boards are introduced. Compuserve offers real-time chat. Many bulletin board systems (BBS) are launched offering online access to other users. The Institute of Electrical and Electronics Engineers (IEEE) publishes the first standards for Ethernet.

**1981**  IBM releases its personal computer (PC), which runs on Microsoft's MS-DOS.

**1982**  The Commodore 64 is released. Having sold more than 22 million units by 1993, it is the world's best-selling single computer. Transmission Control Protocol and Internet Protocol (TCP/IP) is established as ARPANET's standard.

**1983**  Apple's Lisa is released; it is the first personal computer that runs a graphical user interface. Compaq releases the first clone of an IBM PC.

**1984**  Phillips and Sony introduce the CD-ROM. Optical storage begins to replace magnetic storage for some applications.

**1989**   Tim Berners-Lee lays the groundwork ideas of the World Wide Web.

**1990**   The first Ethernet switch is released by Kalpana.

**1993**   Atmel releases a microcontroller with flash memory.

**1997**   IBM produces a copper-based chip, which requires less electricity and generates less heat to allow for up to 200 million transistors on a single chip.

**1998**   Bell Labs introduces the concept of a plastic transistor that is printable.

**2003**   Intel introduces the PXA800F, a microprocessor that combines both computer processes and cellular processes on a single piece of silicon.

**2009**   The first touch-screen flexible e-paper was developed by a team from Arizona State University and E-Ink.

**2009**   Professor Derek Lovely and a team at University of Massachusetts at Amherst engineer a strain of Geobacter bacteria that can produce electric power from mud and wastewater at eight times the normal rate.

**2010**   MIT research team led by Rahul Sarpeshkar develops a new chip that mimics the way the human ear processes signals.

# Mathematics for Electronics

Do you remember sitting in algebra class and wondering, "When will I ever use this?" Now's the time. Electronics gives practical use to many of the math skills you learned in school. In this appendix, we freshen up a few concepts so you're ready to put the basic math of electronics to use.

## Basic Operations and Symbols

Most of the math you use in electronics is simple multiplication, division, and solving simple algebraic questions. Before you start crunching numbers in your calculator take a few minutes to review how to read the symbols in a formula, learn about exponents, review order of operations, and then learn about binary numbers.

### Multiplication

Multiplication can be represented in a variety of ways using different symbols. The following equations represent multiplying X times Y:

X × Y

XY

X • Y

X(Y)

In this book, we generally use the X × Y format.

## Division

Division is usually represented in one of the following ways:

$$X \div Y$$

$$X / Y$$

In this book, we use the X / Y format.

## Scientific Notation

Using scientific notation makes it much easier to solve equations with incredibly large or incredibly small numbers.

To express a large number in scientific notation, follow these steps:

1. Determine the coefficient. The coefficient is the single digit at the beginning of the number followed by a decimal point and then any numbers that precede the zeros. For example, the coefficient of 4,560,000 would be 4.56.

2. Count the number of decimal spaces that follow the first digit. This is the power of 10. Our example is 10 to the sixth power, which is written as $10^6$.

3. Finally, express your number as the coefficient times the power of 10. So our number expressed in scientific notation is $4.56 \times 10^6$.

For extremely small numbers, you can use negative powers of 10:

1. Determine the coefficient. Look for the first significant digit—that is, the first nonzero number. If you are trying to express the number –.00000000056, the coefficient is 5.6.

2. Count the number of decimal spaces that precede and include your first significant digit—in this case, 10 to the negative 10th power, or $10^{-10}$. It is negative because you are moving to the right of the decimal point.

3. Express the number as the coefficient times the power of 10, which is $5.6 \times 10^{-10}$. Notice that it is the power of 10 that is negative, not the coefficient.

## Engineering Notation

In engineering and electronics, most practitioners use engineering notation. Engineering notation sticks to exponents in multiples of 3: $10^{-3}$, $10^0$, $10^3$, and $10^6$. This keeps the numbers in line with the metric prefixes such as milli-, kilo-, and mega-. To express the number 24,000 use $24 \times 10^3$ as engineering notation instead of $2.4 \times 10^4$ in scientific notation.

## Order of Operations

When presented with an equation, it is important to know where to start. This is called order of operations. If you don't follow the proper sequence, you will not get the right answer. If you are feeling a little rusty on these concepts, be sure to double check before proceeding.

First, you solve anything that is in parentheses. This is often called solving parentheses from the inside out. By solving, we mean doing any addition, multiplication, or solving of a fraction. Some equations may have parentheses and brackets surrounding those. Solve the innermost set first.

Second, solve any exponents.

Third, perform any multiplication and division, from left to right.

Fourth, perform any addition and subtraction, from left to right.

## Solving Equations

Electronics involves a lot of equations where you need to use basic algebra to solve an equation to find the value of a variable. There are several steps to follow when you are solving for a variable.

1. First, combine like terms. If you have a set of values of resistance, for instance, you should combine them. $120 \, \Omega + 220 \, \Omega$ should be simplified as $340 \, \Omega$.

2. Next, isolate the variable you wish to solve for. If you want to solve for I (current) in this equation:

   $5 \, V = I / 220 \, \Omega,$

   you need to isolate the variable I. In this case you can multiply each side by 220 to isolate I, as follows:

   $220 \, \Omega \times 5 \, V = I / 220 \, \Omega \times 220$

   $220 \, \Omega \times 5 \, V = I$

   $I = 220 \, \Omega \times 5 \, V$

   $I = 1100 \, A$

   One important note, however, is that when working through electronics equations the various units of measure ($\Omega$, V, and A, for example) aren't variables. We don't want to treat the letters as variables to be solved. You want to leave them in the equation to keep your units straight.

3. Last, substitute your answer into the original equation and check that it works.

## Binary Numbers

Binary is called a base-2 system, which means that each of the place numbers represent a power of 2. (The decimal system is a base-10 numbering system. Binary has 2 digits—0 and 1; decimal has 10—0, 1, 2, 3, 4, 5, 6, 7, 8,and 9.) Working from the right, the first place is held by $2^0$, which is $2 \times 0$, so 0. The next place working from right to left is $2^1$, which is the identity of 2, so just 2. Next, $2^2$, which is $2 \times 2$, so 4. $2^3$, which is $2 \times 2 \times 2$, so 8. This continues infinitely with each place moving to the left representing a further power of 2.

The binary system was used in several ancient traditions, including the Indian; the Chinese (yin as 0 and yang as 1); the Ifà tradition of West Africa; and in medieval geomancy (divination through the use of rocks, sticks, and sand) in Europe, the Middle East, and Africa. In each system, binary numbers could serve as symbols to represent a range of values, and each was formalized so that practitioners could have a uniform way of reading the symbols.

Each binary digit, called a *bit*, is a power of 2. To represent the number 1 in the decimal system as a binary number, you would write 01, the number 2 in decimal is 10 in binary.

One easy approach to convert decimal numbers to binary is to look at a chart, such as the one included in this section, that has the values of the powers of 2. Look for the largest power of 2 that will go into that number without going below 0, put a 1 in that place, then with the remainder go to the next power of 2, and see if it is large enough to subtract that number. If it is, enter a 1 in that place, if not, enter a 0, and then do the same until you have no remainder, entering 0s in any places between your 1s and the decimal point.

For example, let's convert the number 9 into binary. $2^3$ or 8 is the largest power of two that can be subtracted. So we know that the first 1 we can enter is in the fourth place from the right. 1XXX. 9 – 8 = 1, so our remainder is 1. $2^2$ or 4 is the next power of two. As you cannot subtract 4 from 1, you would enter a zero in that place. 10XX. Next, $2^1$ is 2. Again, you cannot subtract 2 from 1, so add another 0. 100X. The last digit would be 1, so the decimal value 9, is 1001 in binary.

There are eight bits in a byte. A kilobyte is $2^{10}$, or 1,024 bytes.

The following table lists some decimal values written in binary form:

| Decimal equivalent | $2^3$ (8) | $2^2$ (4) | $2^1$ (2) | $2^0$ (0) | Binary value |
|---|---|---|---|---|---|
| 0 | 0 | 0 | 0 | 0 | 0000 |
| 1 | 0 | 0 | 0 | 1 | 0001 |
| 2 | 0 | 0 | 1 | 0 | 0010 |
| 3 | 0 | 0 | 1 | 1 | 0011 |
| 4 | 0 | 1 | 0 | 0 | 0100 |
| 5 | 0 | 1 | 0 | 1 | 0101 |
| 6 | 0 | 1 | 1 | 0 | 0110 |
| 7 | 0 | 1 | 1 | 1 | 0111 |
| 8 | 1 | 0 | 0 | 0 | 1000 |
| 9 | 1 | 0 | 0 | 1 | 1001 |
| 10 | 1 | 0 | 1 | 0 | 1010 |
| 11 | 1 | 0 | 1 | 1 | 1011 |
| 12 | 1 | 1 | 0 | 0 | 1100 |
| 13 | 1 | 1 | 0 | 1 | 1101 |
| 14 | 1 | 1 | 1 | 0 | 1110 |
| 15 | 1 | 1 | 1 | 1 | 1111 |

# Careers in Electronics

Now that you have explored the basic concepts in electronics, you might want to explore a career in the field. This appendix gives you a taste of the various levels of electronics education and some of the careers that draw on knowledge of electronics.

## Training for a Career in Electronics

There are four general levels of electronics study, each opening up different career paths.

### The Hobbyist and Curious Self-Taught

General knowledge of electronics and the ability to learn on your own opens many doors. You can teach yourself how to safely repair your own electronics around your home, build and share your projects with groups of other like-minded hobbyists, and even use components to make new inventions.

By understanding electronics, you can be a more educated consumer and share your knowledge and passion with your community. As a journalist or analyst, you can have greater understanding of how things work, enabling you to see the potential in new developments. By keeping up with industry news you can perhaps even use this knowledge to become a better investor by seeing potential in new start-ups or technologies before the rest of the pack catches on.

### Electronics Technician

Electronics technicians train in a community college or technical school environment. They learn professional approaches to manufacturing, testing, repairing, and maintaining equipment from the household consumer level up to the most advanced technologies across many disciplines. The emphasis is on hands-on and experiential learning, but they should also have a solid grasp of the theory behind electronics.

To keep current, electronics technicians should be aware of new developments and technologies through industry publications or further study. Using the same skills and curiosity as the self-taught, an electronics technician can bring a lot of value to almost any sort of technology. Being flexible and curious ensures a long-lasting career.

Electronics technicians should emphasize science and math in their high school programs, with special attention to physics and algebra.

## Electronics Technologists or Engineers

Students who undertake the four years of study leading to a Bachelors of Science in Electrical Engineering, Electronics Engineering, or Computer Engineering work primarily on building a theoretical understanding of the design and operation of electronic devices. This training enables them to better analyze the performance of electronic systems and to design solutions to address the needs of industry. Electrical engineers can also work in sales, project management, or as consultants.

Those who want to pursue a career as an Engineer or Technologist should take advantage of all of the math courses available to them in high school to prepare for the study of math in college. Four years of science is also needed, including chemistry and physics.

## Graduate Level Electronics Engineer

People with a Master's degree or doctorate degree in electronics focus on finding new ways to solve problems and to move the science of electronics forward by developing new technologies.

Students at this level compete against the best talent from around the world. They must have excellent academics and a creative mind.

# Career Fields

You can put your knowledge of electronics to use in the following fields.

## Computers and Information Technology

Electronics and computers go hand in hand. Careers in the field of Information Technology (IT) might involve specializing in computer-aided design, in networking, or in computer support (which requires knowledge of both software and some electronics).

If you want to pursue a career in IT, you should major in either computer science or computer engineering. The field of computer science in most colleges and universities concentrates on the theories of problem solving, algorithms in math, artificial intelligence, and symbols. Computer engineering is the more hardware-oriented field of study. It deals with the components, circuits, and networks linking computers and how they function and communicate.

Knowledge of electronics helps you "think like a computer" so you know both the power and the limitations of a computer's processing ability.

Many high schools offer courses in computer science and IT. You can also gain a lot of knowledge through independent study, courses taught at training centers and community colleges, and by hands-on experimentation. You can gain basic certifications that establish your knowledge of a particular subject to help you establish your proficiency to potential employers.

## Automotive and Transportation

All cars manufactured today are a combination of mechanical and electronic elements. Being a gear-head now requires a basic understanding of the electronics that drive a modern engine. Even if you want to upgrade your car's audio system, you need to know your amps from your watts and be able to read a schematic.

Communication and control systems on boats, ships, and trains are all reliant on electronics, and the field of avionics encompasses the electronics necessary to communicate, navigate, display information, and even fly airplanes, jets, satellites, and spacecraft.

## Biotechnology

The field of biotechnology combines technology and medicine. Whether maintaining or designing monitors, lab equipment, surgical devices, or even bionics that replace organs or structures, electronics are essential to this fast-growing field. Inventors with all levels of training have really expanded what is possible, and their creativity and knowledge can be highly rewarding both in financial terms and in improving the quality or length of life.

## Manufacturing

In manufacturing, electronics are used not only in the final product but also in the many processes used to create that product and get it to market. Whether using robotics to perform a highly detailed manufacturing step or to track inventory in a plant, electronics allow for more efficient methods of producing more advanced products.

## Legal

To act as a patent attorney, one needs to have a technical grounding in a particular field and be trained as an attorney. People working in this specialized field create, protect, and defend the rights of those who hold patents on their intellectual property.

A law degree is not enough to be qualified as a patent attorney. One most also hold at least a Bachelor's degree in a particular area of specialization. Because of the specific educational requirements to be a patent lawyer, there is significant demand for their services.

## Entertainment and Broadcast Media

Whether filling an auditorium with the sound of a few musicians or broadcasting breaking news from a remote location, knowledge of electronics is essential in almost every aspect of entertainment and broadcast media. Using new tools to generate new sounds, music producers are often experts in using electronics to manipulate sound waves. The process of broadcasting a satellite signal from a television studio or a city square thousands of miles away relies on the skills of trained technicians. Advances in electronics and computer processing make movie monsters come alive and 3-D images tower above audiences.

# Resources

One of the most enjoyable aspects of working with electronics is being part of a community of electronics hobbyists. The new wave of electronics aficionados call themselves "Makers," and you will find communities of Makers both online and in your own backyard. Following are some of the sites where Makers gather to find parts, learn about electronics theory, trade projects, and share advice.

## Shopping Resources

Here are places where we purchased items in this book.

**General stores for electronics:**

Sparkfun Electronics
www.sparkfun.com/

Sparkfun is more than a site that sells electronics components; it describes itself as a site that believes in "Sharing Ingenuity." The site includes a blog, informational videos, and comments from other customers on specific products in their forums.

RadioShack
www.radioshack.com

RadioShack was the original corner electronics store that served a couple of generations of electronics hobbyists. Now it has shifted away from a primary focus on electronic components to a more consumer technology store, but it still has cases with drawer after drawer of electronics goodies. RadioShack is great for a quick run for a forgotten part and a one-stop shop for the basics to get you started.

Jameco Electronics
www.jameco.com

Jameco is a great resource for electronics for the hobbyist, the student, and the professional. It offers a very broad range of products including brand names, as well as more affordable components.

### More specialized resources:

Extech Instruments
www.extech.com

Extech is a manufacturer and supplier of test and measurement instruments including the DMM we use in all of our projects. Extech instruments are tools that will serve you for years, from the first days of your electronics education on to your professional career.

Tamiya USA
www.tamiyausa.com

If you want to explore more of the radio control (RC) hobby branch of electronics, Tamiya has all the controllers, models, parts, and RC vehicles you could need. Sparkfun recommends their wheel assemblies and axles for building robots.

### Microcontroller sources:

Secret Labs
www.netduino.com

This is the site for the manufacturer of the microcontroller we use. It includes instructions, tutorials, and a great support forum that highlights member projects and is helpful for anyone who runs into a problem.

Arduino Microcontrollers
www.arduino.cc/en/Main/Hardware

The Arduino controller is another very popular microcontroller and community.

Parallax
www.parallax.com

Microcontroller that can be programmed using the BASIC programming language.

## Information Resources

San Jose State University's Dr. Buff Furman's Mechatronics Portal
www.engr.sjsu.edu/bjfurman/courses/ME106/mechatronicstutorials.htm

Dr. Furman links to many online resources; from the basics to much more advance resources.

Microsoft Robotics
www.microsoft.com/robotics/

A portal site for Microsoft's Robotics Developer Studio R3; it includes case studies, tutorials, and links to robotics sources.

Lady Ada
www.ladyada.net

The site of inventor/artist Limor Fried ("Lady Ada"), it includes her portfolio, blog, resources, and research. She is both an inventor and an agitator but always in the name of science.

Data Sheet Catalog
www.datasheetcatalog.com

This site offers data sheets for just about everything.

*Make:* Online
www.makezine.com, www.makershed.com, and www.makerfaire.com

*Make* magazine's tagline is "technology on your time." This magazine both inspires and reflects a renaissance of hands-on invention. The magazine's site offers podcasts related to the material covered in the print version, and the companion store, Makershed, offers project kits, tools, and books. Maker Faire is the in-person gathering of Makers and vendors now held in three cities annually and other smaller Mini Maker Faires.

Society of Robots
www.societyofrobots.com

Roboticist John Palmisano's site has materials for beginners through much more advanced robot designers.

## Tech and Other Interesting Resources

Think Geek
www.thinkgeek.com

A store for everything geeky.

Slashdot.org

As its tagline says, "News for nerds. Stuff that matters." Slashdot provides links to stories in science, engineering, and other nerdy concerns but its the interaction between the community of commenters that makes this site so beloved.

Gizmodo
www.gizmodo.com

Electronics and gadgets news including coverage of microcontrollers and other hands-on electronics.

Engadget
www.engadget.com

Electronic gadgets news, reviews, and electronics technology discussions.

## Programming the Netduino on an Apple Computer

At the writing of this book there is no way to write and compile C# for .NET using Apple Mac OS. There is a lot of work being done on providing compatibility but it may be some time.

One way to write and compile C# for .NET with Mac OS is to run a virtual Windows system on your Apple computer.

To do this, download a free copy of Virtual Box for OS X hosts (Intel Macs only). Virtual Box is a free program for OS X systems and is available at: www.virtualbox.org/wiki/Downloads.

You'll also need to purchase a copy of one of the compatible Windows operating systems. You can use Windows XP, Vista, or 7. You will need to read the instructions on setting up the virtual Windows system on your Apple computer. You can find the instructions here: www.virtualbox.org/manual/UserManual.html.

After you have your virtual Windows system up and running, follow the instructions in Chapter 18 to set up and install all of the programs and tools for programming the Netduino.

# Index

# X–Y

# Z

# CHECK OUT THESE BEST-SELLERS

More than 450 titles available at booksellers and online retailers everywhere!

978-1-59257-115-4

978-1-59257-900-6

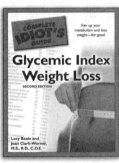

978-1-59257-855-9

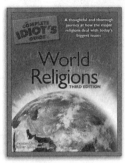

978-1-59257-222-9

978-1-59257-957-0

978-1-59257-785-9

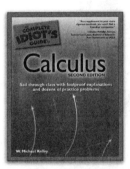

978-1-59257-471-1

978-1-59257-483-4

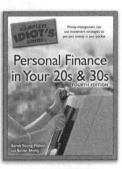

978-1-59257-883-2

978-1-59257-966-2

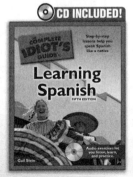

978-1-59257-908-2

978-1-59257-786-6

978-1-59257-954-9

978-1-59257-437-7

978-1-59257-888-7

ALPHA    idiotsguides.com

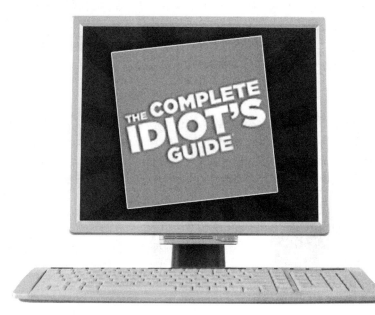